Optimization with LINGO-18
Problems and Applications

Neha Gupta
Assistant Professor, Amity School of Business
Amity University, Uttar Pradesh, India

Irfan Ali
Assistant Professor, Department of Statistics & Operations Research
Aligarh Muslim University, Aligarh, India

CRC Press
Taylor & Francis Group
Boca Raton London New York

CRC Press is an imprint of the
Taylor & Francis Group, an **informa** business

A SCIENCE PUBLISHERS BOOK

First edition published 2021
by CRC Press
6000 Broken Sound Parkway NW, Suite 300, Boca Raton, FL 33487-2742

and by CRC Press
2 Park Square, Milton Park, Abingdon, Oxon, OX14 4RN

© 2021 Taylor & Francis Group, LLC

CRC Press is an imprint of Taylor & Francis Group, LLC

Reasonable efforts have been made to publish reliable data and information, but the author and publisher cannot assume responsibility for the validity of all materials or the consequences of their use. The authors and publishers have attempted to trace the copyright holders of all material reproduced in this publication and apologize to copyright holders if permission to publish in this form has not been obtained. If any copyright material has not been acknowledged please write and let us know so we may rectify in any future reprint.

Except as permitted under U.S. Copyright Law, no part of this book may be reprinted, reproduced, transmitted, or utilized in any form by any electronic, mechanical, or other means, now known or hereafter invented, including photocopying, microfilming, and recording, or in any information storage or retrieval system, without written permission from the publishers.

For permission to photocopy or use material electronically from this work, access www.copyright.com or contact the Copyright Clearance Center, Inc. (CCC), 222 Rosewood Drive, Danvers, MA 01923, 978-750-8400. For works that are not available on CCC please contact mpkbookspermissions@tandf.co.uk

Trademark notice: Product or corporate names may be trademarks or registered trademarks and are used only for identification and explanation without intent to infringe.

ISBN: 978-0-367-50122-8 (hbk)
ISBN: 978-0-367-50123-5 (pbk)
ISBN: 978-1-003-04889-3 (ebk)

Typeset in Times New Roman
by Radiant Productions

Preface

A lot of work in the field of optimization has been done in the form of research papers, articles and books. This book presents the concepts and applications of optimization straightforwardly. Every concept started with a more developed resolution and explained by examples for better understanding, and the way to write down the LINGO-18 codes for optimization problems is given in detail. The focus of this book is to provide profound study materials on optimization to the reader so that it enables them to understand the usefulness of optimization in various disciplines, including management, engineering, economics and many other application areas.

All book chapters are written in a reader-friendly manner and theoretical concepts with LINGO-18 software applications have been discussed. Chapters of this book provide a basic understanding of various applications of optimization techniques in decision-making problems. After reading this book, the readers will understand the theory and formulations of decision-making problems and their LINGO-18 codes for several optimization problems. Section I: Essential Mathematics and Software, briefly covered the elementary knowledge of calculus, linear algebra and LINGO software. Section II: Optimization, presents the fundamental concept of optimization, linear optimization, non-linear optimization, uncertain optimization, and multi-objective optimization in simple language using examples with their LINGO-18 codes. The writing codes for LINGO-18 software are shown step by step so that readers can easily understand and create the codes for their problems.

This book is designed for academicians, practitioners, researchers, and postgraduate students. After reading this book, the readers will understand the theory and formulations of decision-making problems and their solution obtained by appropriate optimization techniques using LINGO software. The book's contents will be a centre of attraction for the intended audience because of its writing, explanations, examples, and software usage. This book will get the fundamental concepts, problems, and applications of optimization under one umbrella.

For further improvement, reader's suggestions on the book content are welcome.

Contents

Preface iii

Authors' Biographies xi

Section I: Essential Mathematics and Software

1. Basic Calculus   2

   1.1   Distance Formula for Points in the Plane   2
   1.2   Circle Centered at the Origin   2
   1.3   Straight Lines   2
   1.4   Angle of Inclination   3
   1.5   Parallel and Perpendicular Lines   3
   1.6   Function   3
   1.7   Domain   4
   1.8   Functions Graphs   5
   1.9   Composite Functions   5
   1.10 Even and Odd Functions-Symmetry   5
   1.11 Piecewise Defined Functions   6
   1.12 Shifting Graphs   6
   1.13 Average Rate of Change and Secant Lines   8
   1.14 Continuity   8
          1.14.1 Lipschitz Continuity   9
   1.15 Slope of a Curvilinear Function   10
   1.16 Vertical Tangent   11
   1.17 Derivative of the Function   11
   1.18 Differentiable in Interval   12
   1.19 Extreme Values of Functions   13

1.20 Local and Absolute (Global) Extrema — 13
1.21 Increasing Functions and Decreasing Functions — 15
1.22 Tests Based on First Derivative — 16
1.23 Concavity — 17
1.24 Inflection Point — 18
1.25 Linearization Approximation — 20

2. Matrix and Determinant Algebra — 22

2.1 Matrix — 22
2.2 Types of Matrix — 23
2.3 Algebra of Matrix — 23
    2.3.1 Addition and Subtraction of Matrix — 23
    2.3.2 Multiplication by a Scalar — 23
    2.3.3 Matrix Multiplication — 23
    2.3.4 Laws of Algebra — 24
2.4 Transpose of Matrix — 24
2.5 Trace of Matrix — 25
2.6 Non-Singular Matrix — 25
2.7 Inverse of a Matrix — 25
2.8 Orthogonal Matrix — 26
2.9 Rank of Matrix — 26
2.10 Determinants and Non-Singularity — 26
2.11 Three Order Determinants — 28
2.12 Linear Equations — 30
2.13 Vector Algebra — 31
    2.13.1 Metric Spaces — 31
    2.13.2 Normed Linear Spaces — 32
    2.13.3 Norms — 33
    2.13.4 Forbenlus Norm — 34
    2.13.5 Eigenvalues Vectors — 34

3. LINGO-18: Optimization Software — 36

3.1 Introduction — 36
    3.1.1 Key Benefits of LINGO — 36
    3.1.2 Installing LINGO — 37
    3.1.3 Window Types in LINGO — 37
    3.1.4 Model Creation and Solution in LINGO — 38
    3.1.5 Logical Operators and Set Looping Functions — 42
    3.1.6 Variable Domain Functions — 43
    3.1.7 Shortcut Keys — 44

Section II: Optimization

4. Introduction to Optimization — 46

- 4.1 Introduction — 46
  - 4.1.1 Unconstrained vs Constrained — 46
  - 4.1.2 Linear vs Non-Linear — 47
- 4.2 Basic Definitions — 47
  - 4.2.1 Affine Sets — 47
  - 4.2.2 Convex Set — 48
  - 4.2.3 Convex Function — 48
  - 4.2.4 Concave Function — 49
- 4.3 Convex and Concave Properties — 51
- 4.4 Convex Optimization Problems — 52
- 4.5 Concave Maximization Problem — 52
  - 4.5.1 Quasi Concave & Quasi Convex Function — 53
    - 4.5.1.1 Properties of Quasi Concave Functions — 54
- 4.6 Convexity of Smooth Function — 56
  - 4.6.1 Quasi Concavity of Smooth Function — 57
  - 4.6.2 Pseudo Concave Function — 57
  - 4.6.3 Pseudo Convex Function — 57
  - 4.6.4 Cone — 58
  - 4.6.5 Hyperplanes and Half Spaces — 58
- 4.7 Univariate Optimization Problems — 59
  - 4.7.1 Optimality Condition — 59
- 4.8 Multivariate Optimization Problems — 61
  - 4.8.1 Optimality Condition — 61
- 4.9 Gradient Vector of $f(x)$ — 65
  - 4.9.1 Hessian Matrix of $f(x)$ — 65

5. Linear Optimization Problems — 68

- 5.1 Introduction — 68
- 5.2 General Form of LP Model — 69
  - 5.2.1 Components of LP Model — 69
  - 5.2.2 Assumptions — 69
  - 5.2.3 General Mathematical Model — 70
  - 5.2.4 Properties of Solution to an LPP — 72
- 5.3 Formulation of an LPP — 72
  - 5.3.1 Product Mix Problem — 72
  - 5.3.2 Resource Allocation Problem — 74
  - 5.3.3 Diet Problem — 74
  - 5.3.4 The Capital Budgeting Problem — 75
  - 5.3.5 Assignment Problem — 76
  - 5.3.6 Transportation Problem — 77

5.4	Integer Programming	77
5.5	Graphical Solution of LPP	78
5.6	Theory of Simplex Method	83
	5.6.1 Standard Form	83
	5.6.2 Computational Procedure of Simplex Method	84
	5.6.3 Maximization Problem	86
	5.6.4 Degeneracy and Cycling in LPP	88
	5.6.5 Artificial Basis Technique	88
	5.6.6 Minimization Problem	89
5.7	Solving LPP using LINGO-18	91
	5.7.1 Product Mix Profit Maximization Problem	92
	5.7.2 Product Sales through Advertisement Problem	93
	5.7.3 Profit Maximization Problem	94
	5.7.4 Total Production Cost Minimization Problem	95
	5.7.5 Product Manufacturing Problem	97
	5.7.6 Product Manufacturing Problem	97
	5.7.7 Production Cost Minimization Problem	98
	5.7.8 Profit Maximization Problem	100
	5.7.9 Cost Minimization Problem	101
5.8	Concept of Duality in LPP	102
	5.8.1 Conversion of Primal to Dual	103
	5.8.2 Importance of Duality Concepts	106
	5.8.3 Properties of Primal-dual LPPs	109
	5.8.4 Economic Interpretation of Duality	109
	5.8.5 Dual Simplex	111

6. Non-Linear Optimization Problems   **115**

6.1	Introduction	115
	6.1.1 Basic Definitions	115
	6.1.2 Some Properties	117
6.2	Lagrange Multiplier	117
6.3	Kuhn-Tucker Conditions	118
6.4	Solution of Non-Linear Optimization Problems using LINGO-18	119
6.5	Quadratic Programming	122
	6.5.1 Wolf's Method	123
	6.5.2 Beale's Method	126
	6.5.3 Algorithm	126
6.6	Solution of Quadratic Programming Problems using LINGO-18	129
6.7	Convex Programming	131
	6.7.1 A Feasible Direction	131
	6.7.2 A Usable Feasible Direction	131
	6.7.3 An Outline of the Methods of Feasible Directions	132

		6.7.4	Rosen's Gradient Projection Method	132
		6.7.5	Kelly's Method	135
	6.8	Solution of Convex Programming Problems using LINGO-18		138

7. Optimization Under Uncertainty   141

    7.1   Introduction   141
    7.2   Fuzzy Optimization   141
         7.2.1   Introduction   141
         7.2.2   Operations of Fuzzy Sets   143
         7.2.3   Cardinality of Fuzzy Set   145
               7.2.3.1   Scalar Cardinality   145
               7.2.3.2   Relative Cardinality   145
               7.2.3.3 α-cut Set   149
               7.2.3.4   Strong α-cut   149
               7.2.3.5   Level Set   149
    7.3   Fuzzy Numbers   151
         7.3.1   Triangular Fuzzy Number   151
         7.3.2   Trapezoidal Fuzzy Number   151
    7.4   Defuzzification Methods   152
         7.4.1   Center of Sums Method   152
         7.4.2   Center of Gravity Method   154
         7.4.3 α-cut Method   155
    7.5   Solving Optimization Problem with Fuzzy Numbers using LINGO-18   156
         7.5.1   Case I: When Coefficients in Objective Function are Fuzzy Numbers   157
         7.5.2   Case II: When Constraint Coefficients are Fuzzy Numbers   158
         7.5.3   Case III: When RHS Parameters are Fuzzy Numbers   159
    7.6   Stochastic Optimization Problem   161
         7.6.1   Situation I: Parameters in Objective Function C_j are Random Variables   161
         7.6.2   Situation II: Availability/Requirement Vector b_i are Probabilistic   162
         7.6.3   Situation III: When a_{ij} are Random Variables   164
         7.6.4   Situation IV: General Case—All Parameters are Random Variables   165
    7.7   Some Numerical Examples using LINGO-18   167
    7.8   Interval Optimization Problem   170
         7.8.1   Introduction   170
         7.8.2   Interval Arithmetic   170
         7.8.3   Formulation of Interval Optimization Problem   171
         7.8.4   Algorithm   172
         7.8.5   Numerical Example using LINGO-18   173

8. Multi-Objective Optimization — 176

- 8.1 Introduction — 176
  - 8.1.1 Pareto Optimal Solution — 178
  - 8.1.2 Ideal Point — 179
  - 8.1.3 Anti-ideal Point — 180
  - 8.1.4 Nadir Point — 180
  - 8.1.5 Pareto Frontier — 181
- 8.2 Multi-Objective Optimization Techniques — 181
  - 8.2.1 Weighted Technique — 181
  - 8.2.2 Goal Programming Technique — 182
    - 8.2.2.1 Algorithm of Goal Programming Technique — 183
    - 8.2.2.2 Another Standard Form of GP — 184
  - 8.2.3 Lexicographic Goal Programming Technique — 185
  - 8.2.4 ε-Constraints Technique — 186
  - 8.2.5 Fuzzy Goal Programming Technique — 186
  - 8.2.6 Fuzzy Goal Programming with Tolerance — 188
- 8.3 Numerical Example using LINGO-18 — 188
  - 8.3.1 Solution through Weighted Technique — 189
  - 8.3.2 Solution through Goal Programming Technique — 190
  - 8.3.3 Solution through Lexicographic Goal Programming Technique — 191
  - 8.3.4 Solution through ε-Constraints Technique — 192
  - 8.3.5 Solution through Goal Programming Technique — 193
  - 8.3.6 Solution through Fuzzy Goal Programming Technique — 195
- 8.4 Multi-Objective Optimization Problem with Fuzzy Numbers — 196
  - 8.4.1 Algorithm — 197
- 8.5 Numerical Example using LINGO-18 — 197

9. Applications of Optimization — 200

- 9.1 Optimization Problems in Finance — 200
- 9.2 Optimization Problems in Marketing — 206
- 9.3 Optimization Problems in Human Resource Management — 212
- 9.4 Vendor Selection Optimization Problem — 214
- 9.5 Diet Optimization Problem — 217
- 9.6 Operations Management Optimization Problems — 221
- 9.7 Transportation Optimization Problem — 225
- 9.8 Assignment Optimization Problem — 227

References — 230

Index — 233

Authors' Biographies

Dr. Neha Gupta is currently working as Assistant Professor in the Amity School of Business, Amity University Uttar Pradesh, Noida (India). She started her career by working with SSLD Varshney Group of Institutions and Shree Guru Gobind Singh Tricentenary University, with more than of 5 years of rich experience in academics and research.

Dr. Gupta has published more than 25 research papers in journals of national and international repute including publishers like Taylor & Francis and Springer. She is also a member in the Editorial Board of International Journal of Mathematics and Systems Science, International Journal of Data Mining, Modelling & Management-Inderscience, and Editorial Review Board of OPSEARCH, Journal of Statistical Computation and Simulation, and International Journal of Management Science and Engineering Management. She is currently editing one book to be published by Springer Nature.

She is a Gold Medalist and a life member of Operational Research Society of India. She has presented papers at various National & International conferences and has won best research paper award. Her research interest includes Optimization, Decision-Making, Transportation, Operations Management, Supply Chain Management, Mathematical Programming, and Data analysis using R, SPSS, LINGO, MATLAB, etc.

Irfan Ali received the B.Sc., M.Sc., M.Phil., and Ph.D. degrees from Aligarh Muslim University. He is currently working as an Assistant Professor with the Department of Statistics and Operations Research, Aligarh Muslim University. His research interests include applied statistics, survey sampling, supply chain networks and management, mathematical programming, and multiobjective optimization. He has supervised M.Sc., M.Phil., and Ph.D. students in operations research. He has completed a research project UGC–Start-Up Grant Project, UGC, New Delhi, India. He has published more than 75 research articles in reputed journals and serves as a Reviewer for several journals. He is currently editing two books to be published by Taylor & Francis and Springer Nature.

He is a Lifetime Member of various professional societies: Operational Research Society of India, Indian Society for Probability and Statistics, Indian Mathematical Society, and The Indian Science Congress Association. He delivered invited talks in several universities and Institutions. He was a recipient of the Post Graduate Merit Scholarship Award during M.Sc. (statistics) in 2008 and the UGC-BSR Scholarship awarded during the Ph.D. (statistics) programs in 2013. He also serves as an Associate Editor for some journals.

ESSENTIAL MATHEMATICS AND SOFTWARE

I

Chapter 1

Basic Calculus

In this chapter, the preliminaries of calculus are briefly discussed.

1.1 Distance Formula for Points in the Plane

The distance d between two points $P(x_1, y_1)$ and $Q(x_2, y_2)$ on a plane is given as

$$d = \sqrt{(x_2 - x_1)^2 + (y_2 - y_1)^2}$$

1.2 Circle Centered at the Origin

1. Let the equation $x^2 + y^2 = a^2$, $a > 0$ represents all points $P(x, y)$ whose distance from the origin is $d = a$. These points lie on the circle of radius a centered at the origin.

2. All points $P(x, y)$ whose coordinates satisfy the inequality $x^2 + y^2 \leq a^2$, $a > 0$ have a distance $d \leq a$ from the origin.

1.3 Straight Lines

The increment between two points $P(x_1, y_1)$ and $Q(x_2, y_2)$ in the plane are $\Delta x = x_2 - x_1$ and $\Delta y = y_2 - y_1$ called as the *run* and the *rise* respectively. Whereas $\frac{\Delta y}{\Delta x} = \frac{y_2 - y_1}{x_2 - x_1} = m$ has the same value for every choice of the two points $P(x_1, y_1)$ and $Q(x_2, y_2)$ on the line where m is the slope of the non-vertical line PQ.

1.4 Angle of Inclination

The inclination of a horizontal line is 0° and the inclination of a vertical line is 90°. If θ is the inclination of line, then $0^o \leq \theta \leq 180^o$.

The relation between the slope m of a non vertical line and the line's angle of inclination θ is $m = \tan\theta = \dfrac{\Delta y}{\Delta x}$. The slope of a non-vertical line is the tangent of its angle of inclination (see Fig. 1.1).

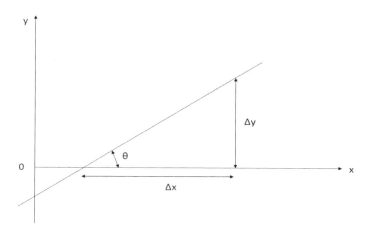

Figure 1.1: Angle of Inclination.

1.5 Parallel and Perpendicular Lines

Parallel lines have equal angles of inclination. Hence, they have the same slope if they are not vertical. Conversely, lines with equal slopes have equal angles of inclination and so are parallel.

If two non vertical lines L_1 and L_2 are perpendicular then their slopes m_1 and m_2 satisfy $m_1 \times m_2 = -1 \Rightarrow m_1 = -\dfrac{1}{m_2}$ or $m_2 = -\dfrac{1}{m_1}$. So, each slope is the negative reciprocal of the other.

The equation for a non vertical straight line L if we know its slope m and the coordinates of points $P(x_1, y_1)$ and $Q(x, y)$ on it, then $m = \dfrac{y - y_1}{x - x_1}$, so that $y = y_1 + m(x - x_1)$.

1.6 Function

Let $y = f(x)$ be a function where value of variable y depends on the value of variable x. Since the value of y is completely determined by the value of x, then y is a function of x.

Definition 1.1 A function maps from a domain set D to a set containing the range R assigning a unique element from R to each element of D (see Fig. 1.2).

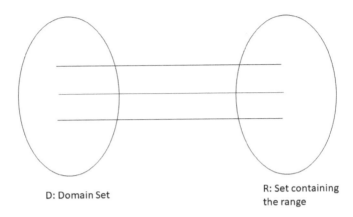

D: Domain Set

R: Set containing the range

Figure 1.2: Function Mapping.

1.7 Domain

Let function $y = f(x)$ where the domain is not stated explicitly; the domain is assumed to be the largest set of x values for which the function y gives real values. This domain is known as the function's natural domain.

Example 1 *The domain of the function $y = f(x) = x^2$ is the entire set of real numbers. The function gives a real y value for every real number x. If we want to restrict the domain to values of x, that is, $y = x^2$, $x \geq 3$ it means changing the domain which changes the range. The range of $y = f(x) = x^2$ is $[0, \infty)$. The range $y = x^2$, $x \geq 3$ is $[9, \infty)$.*

Example 2 *The domain of the function $y = f(x) = \sqrt{1-x^2}$ is $[-1,1]$ and range $[0,1]$. That is function gives a real y value for every x in the closed interval from -1 to 1. Beyond this domain, $1-x^2$ is negative and therefore $\sqrt{1-x^2}$ is not a real number. The values of $1-x^2$ vary from 0 to 1 on the domain $[-1,1]$.*

Example 3 *The domain of the function $y = f(x) = \frac{1}{x}$ is $(-\infty, 0) \bigcup (0, \infty)$ and range $(-\infty, 0) \bigcup (0, \infty)$. That is function gives a real y value for every x, except $x=0$, since we cannot divide any number by zero. The range of the function is the set of all nonzero real numbers.*

Example 4 *The domain of the function $y = f(x) = \sqrt{9-x}$ is $(-\infty, 9]$ and range $[0, \infty)$. That is, $9-x \geq 0$ or $x \leq 9$. The function is real for all $x \leq 9$.*

Basic Calculus ■ 5

1.8 Functions Graphs

Every curve which we draw cannot be the graph of a function. For example, a circle cannot be a function graph since some vertical lines can intersect the circle twice.

As we know that a function f can have only one value $f(x)$ for each x in its domain so that no vertical line can intersect the graph of a function more than once. Let us suppose if a is in the domain of a function f; then the vertical line $x = a$ will intersect the graph of f in the single point $(a, f(a))$, which is not valid in case of a circle.

1.9 Composite Functions

If f and g are two different functions, the composite function $f \circ g$ is defined as $(f \circ g)(x) = f(g(x))$. According to the definition, two functions can be composed when the range of the first function lies in the domain of the second function.

Example 5 *If $f(x) = \sqrt{x}$ and $g(x) = x + 3$, then the following composite functions holds:*

1. $(f \circ g)(x) = f(g(x)) = \sqrt{g(x)} = \sqrt{x+3}$ *in domain* $D = [-3, \infty)$
2. $(g \circ f)(x) = g(f(x)) = f(x) + 3 = \sqrt{x} + 3$ *in domain* $D = [0, \infty)$
3. $(f \circ f)(x) = f(f(x)) = \sqrt{f(x)} = \sqrt{\sqrt{x}}$ *in domain* $D = [0, \infty)$
4. $(g \circ g)(x) = g(g(x)) = g(x) + 3 = (x+3) + 3 = x + 6$ *in domain* $D = \mathbb{R}$ *or* $(-\infty, \infty)$

1.10 Even and Odd Functions-Symmetry

A function $y = f(x)$ is even, if $f(-x) = f(x)$ for each value of x in the domain of f. Note that, x and $-x$ must be in the domain of f. For example, $f(x) = x^2$ is even since $f(-x) = (-x)^2 = x^2 = f(x)$ (see Fig. 1.3).

The graph of an even function $y = f(x)$ is symmetric about the y-axis. Since $f(-x) = f(x)$, the point (x, y) lies on the graph if and only if the point $(-x, y)$ also lies on the graph. Once the graph on one side of the $y - axis$ is known, then other side of the graph is automatically known.

A function $y = f(x)$ is odd if $f(-x) = -f(x)$ for each value of x in the domain of f. Note that, x and -x must be in the domain of f. For example, $f(x) = x^3$ is odd since $f(-x) = (-x)^3 = -x^3 = -f(x)$.

The graph of an odd function $y = f(x)$ is also symmetric about the $y - axis$. Since $f(-x) = -f(x)$, the point (x, y) lies on the graph if and only if the point $(-x, -y)$ lies on the graph. Once we know the graph on one side of the $y - axis$, the other side is automatically known to us.

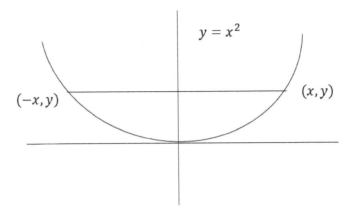

Figure 1.3: Symmetry Curve.

1.11 Piecewise Defined Functions

Sometimes a function is defined with formulas on different parts of its domain (see Fig. 1.4). For example, an absolute value function is defined as

$$y = |x|, \text{ where } |x| = \begin{cases} x, & \text{if } x \geq 0 \\ -x, & \text{if } x < 0 \end{cases}$$

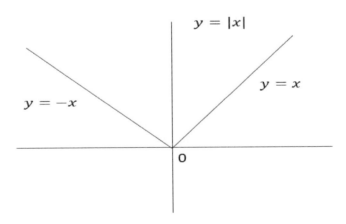

Figure 1.4: Absolute Valued Function.

1.12 Shifting Graphs

Let there be a function $y = f(x)$; to shift the graph straight up, add a positive constant in the equation. And to shift the graph straight down, add a negative constant in the equation.

Example 6 Let there be a function $y = x^2$; to shift the graph straight up by two units; add 2 units in the equation: $y = x^2 + 2$. Similarly, to shift the graph straight down by two units, add -2 units in the equation: $y = x^2 - 2$.

Example 7 Let there be a function $y = x^2$; to shift the graph two units to the left, add 2 units in the equation, that is, $y = (x+2)^2$. Similarly, to shift the graph two units to the right add -2 units in the equation, that is, $y = (x-2)^2$.

For vertical shifts use the formula, $y = f(x) + k$, shift the graph up k units, where $k > 0$. Shift the graph down k units where $k < 0$.

For horizontal shifts use the formula $y = f(x - h)$ shift the graph to the right by h units, where $h > 0$. Shift the graph to the left by h units where $h < 0$.

Example 8 *Equations of circles*
A circle of radius a centred at the origin has equation $x^2 + y^2 = a^2$. If we shift the circle centre from origin to the point (h,k), its equation now becomes $(x-h)^2 + (y-k)^2 = a^2$.
Interior and Exterior Circles:
The interior points that lie inside the circle $(x-h)^2 + (y-k)^2 = a^2$ are the points less than a units from (h,k), and satisfy the inequality $(x-h)^2 + (y-k)^2 < a^2$ (see Fig. 1.5). The circle exterior consists of the points that lie more than a units from (h, k), and satisfy the inequality $(x-h)^2 + (y-k)^2 > a^2$.

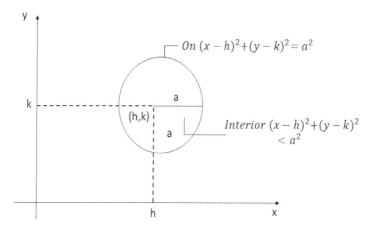

Figure 1.5: Interior and Exterior Circles.

Example 9 *Parabolic Graphs*
The graph of an equation $y = ax^2$ is a parabola whose axis of symmetry is the y-axis. The parabola vertex lies at the origin. The parabola opens upward if $a > 0$

and downward if $a < 0$. The larger the value of a, the narrower the parabola and vice versa.

The graph of an equation $x = ay^2$ is a parabola whose axis of symmetry is the x-axis. The parabola vertex lies at the origin. The parabola opens to the right if $a > 0$ and to the left if $a < 0$. The larger the value of a the narrower the parabola, and vice versa.

Example 10 The Quadratic Equation $y = ax^2 + bx + c$, $a \neq 0$
If we shift the parabola $y = ax^2$ horizontally, that is, $y = a(x-h)^2$; further to shift it vertically as well, we have $y - k = a(x-h)^2$. The combined shifts place the vertex at the point (h,k) and the axis along the line $x = h$. Finally, after rearranging the terms the resultant equation is $y = ax^2 + bx + c$, $a \neq 0$, the curve has same shape and orientation as the curve $y = ax^2$. The graph of the equation $y = ax^2 + bx + c$, $a \neq 0$ is a parabola and it is open upward if $a > 0$ and downward if $a < 0$. The axis is the line $x = -\frac{b}{2a}$. The vertex of the parabola is the point where the axis and parabola intersect. The x coordinate is $x = -\frac{b}{2a}$ and y coordinate is calculated after substituting the x coordinate value in the parabola equation.

1.13 Average Rate of Change and Secant Lines

Let there be a function $y = f(x)$. The average rate of change of $y = f(x)$ with respect to x over the interval $[x_1, x_2]$ is $\frac{\Delta y}{\Delta x} = \frac{f(x_2) - f(x_1)}{x_2 - x_1} = \frac{f(x_1 + h) - f(x_1)}{h} =$ slope. Note that, the average rate of change of f over $[x_1, x_2]$ is the slope of the line through the points $P(x_1, f(x_1))$ and $Q(x_2, f(x_2))$. In geometry, a line joining two points of a curve is called a secant to the curve. Thus, the average rate of change of f from x_1 to x_2 is identical with the slope of secant PQ (see Fig. 1.6).

1.14 Continuity

A function f is continuous at an interior point $x = c$ of its domain if $\lim_{x \to c} f(x) = f(c)$. A function f is continuous at a left end point $x = a$ of its domain if $\lim_{x \to a+} f(x) = f(a)$ and continuous at a right end point $x = b$ of its domain if $\lim_{x \to b-} f(x) = f(b)$. In general, a function f is right continuous (continuous from the right) at a point $x = c$ in its domain if $\lim_{x \to c+} f(x) = f(c)$ and left continuous at c if $\lim_{x \to c-} f(x) = f(c)$ (see Fig. (1.7)).

Example 11 Let $f(x) = \sqrt{9 - x^2}$ be continuous at every point of its domain $[-3, 3]$. This includes $x = -3$, where f is right continuous, and $x = 3$, where f is left continuous.

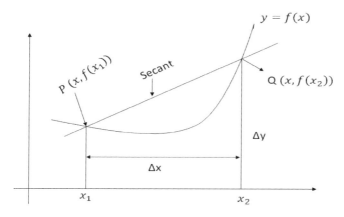

Figure 1.6: Rate of Change and Secant.

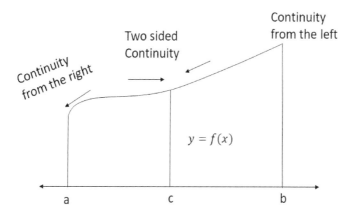

Figure 1.7: Continuity Function.

Example 12 *The function $f(x) = |x|$ is continuous at every value of x. If $x > 0$, we have $f(x) = x$, a polynomial. If $x < 0$, we have $f(x) = -x$, another polynomial. Finally, at the origin, $\lim_{x \to 0} |x| = 0$.*

1.14.1 Lipschitz Continuity

The Lipschitz continuity of a function plays an important role in optimization and convergence analysis. Let x_1 and x_2 two points in the domain of the function $f(x)$, the Lipschitz continuity of function $f(x)$,

$$|f(x_1) - f(x_2)| \leq L|x_1 - x_2| \qquad (1.1)$$

Here in Eqn. (1.1), L is called the Lipschitz constant (or the modulus of uniform continuity), which is independent of x_1 and x_2.

Eqn. (1.1) equivalently to the absolute derivative case, that is

$$\frac{|f(x_1) - f(x_2)|}{|x_1 - x_2|} \leq L \leq \infty \tag{1.2}$$

which limits the rate of change of function.

In accordance to the Eqn. (1.2), a small change in the input variable (or independent variable) can lead to the arbitrarily small change in the output function (or dependent variable). The case, when the Lipschitz constant is sufficiently large, it means a small change in x could lead to a much larger change in $f(x)$, but this lipschtiz estimate is an upper bound and the actual change can be much smaller. It is noted that any function with a finite or bounded first derivative is lipschitz continuous.

Example 13 *Let function $f(x) = 4x^2$ is Lipschitz continuous in the domain $D = (-10, 10)$, we have*

$$|f(x_1) - f(x_2)| = |4x_1^2 - 4x_2^2| = 4|x_1 + x_2||x_1 - x_2|$$

Since $|x_1 + x_2| \leq |x_1| + |x_2| \leq 20$, we have $|f(x_1) - f(x_2)| \leq 80|x_1 - x_2|$, which means that the lipschitz constant is 80. Noted that $L = 80$ is just one lower value. It can be concluded that lipschitz constant does not depend only on x_1 and x_2, it may also depend on the size of the domain D.

1.15 Slope of a Curvilinear Function

The slope of a curvilinear function cannot be constant. It differs at different points on the curve. The slope of a curvilinear function at a given point can be measured by the tangent to the function at that point.

A tangent line is a straight line that touches a curve at only one point. Steps to find a tangent to the curve $y = f(x)$ at point (x_0, y_0) are as follows:

(i) calculate the function $f(x_0)$ and $f(x_0 + h)$.

(ii) calculate the slope, $m = \lim_{h \to 0} \frac{f(x_0 + h) - f(x)}{h}$

(iii) if the limit exist, the tangent line will be found as $y = y_0 + m(x - x_0)$

Example 14 *Let a curvilinear function $y = f(x) = 2x^2$, then the slope of the function is*

$$m = \lim_{h \to 0} \frac{2(x+h)^2 - 2x^2}{h} = \lim_{h \to 0} \frac{2(x^2 + h^2 + 2hx) - 2x^2}{h}$$

$$or \ m = \lim_{h \to 0} \frac{4hx + 2h^2}{h} = \lim_{h \to 0}(4x + 2h)$$

take limit $h \to 0$, then $m = 4x$.

Noted that, the value of the slope m depends on the value of x chosen. At $x = 1$, slope $m = 4$; at $x = 2$, slope $m = 8$ and so on.

Example 15 Find the tangent line for $y = mx + b$ at the point $(x_0, mx_0 + b)$. $f(x_0) = mx_0 + b$ and $f(x_0 + h) = m(x_0 + h) + b$, hence,

$$\lim_{h \to 0} \frac{f(x_0 + h) - f(x_0)}{h} = m$$

Therefore, the tangent line at the point $(x_0, mx_0 + b)$ is $y = mx + b$. It shows that the line $y = mx + b$ is its own tangent at any point $(x_0, mx_0 + b)$.

1.16 Vertical Tangent

Let the curve $y = f(x)$ have a vertical tangent at the point $x = x_0$ where $\lim_{h \to 0} \frac{f(x_0 + h) - f(x_0)}{h} = \infty$ or $-\infty$

Example 16 Let the curve $y = x^{1/3}$, have a vertical tangent at $x = 0$

$$\lim_{h \to 0} \frac{f(0 + h) - f(0)}{h} = \lim_{h \to 0} \frac{h^{1/3} - 0}{h} = \infty$$

1.17 Derivative of the Function

The function f' is the derivative of the function f with respect to the variable x. The f' value at x is $f' = \lim_{h \to 0} \frac{f(x+h) - f(x)}{h}$, provided the limit exists. If f' exists, then f has a derivative at x.

Example 17 Let $f = \frac{x}{x-1}$ and $f(x + h) = \frac{x+h}{x+h-1}$ and hence $f' = \lim_{h \to 0} \frac{-1}{(x+h-1)(x-1)} = \frac{-1}{(x-1)(x-1)}$. The slope will be -1, provided $\frac{-1}{(x-1)^2} = -1 \Rightarrow x = 2$ or $x = 0$.

Example 18 Let $y = \sqrt{x}$, $x > 0$ and $f(x + h) = \sqrt{x+h}$ and hence $f' = \lim_{h \to 0} \frac{-1}{\sqrt{x+h} + \sqrt{x}} = \frac{1}{2\sqrt{x}}$. Note that, function is defined at $x = 0$, but its derivative is not.

The slope of the curve at $x = 9$ is $f' = \frac{dy}{dx}|_{x=9} = \frac{1}{6}$. The tangent is the line passing through the point (9,3) with the slope 1/6.

1.18 Differentiable in Interval

A function $f(x)$ is differentiable in an open interval (finite or infinite), if it has a derivative at each point of the interval. It is differentiable on a closed interval [a,b] if it is differentiable in the interior (a,b) and if the limits

$$\lim_{h \to 0+} \frac{f(a+h)-f(a)}{h} \quad \text{right hand side derivative at a.}$$

$$\lim_{h \to 0-} \frac{f(b+h)-f(b)}{h} \quad \text{left hand side derivative at b.}$$

Example 19 Let the function $y = |x|$. To the right of the origin, differentiation of the function is $\frac{d}{dx}|x| = \frac{d}{dx}(x) = 1$, and to the left of the origin, differentiation of the function is $\frac{d}{dx}|x| = \frac{d}{dx}(-x) = -1$. Note that there can be no derivative at the origin. Since right hand side derivative of $|x|$ at zero is 1, that is, $\lim_{h \to 0+} \frac{|0+h|-|0|}{h} = 1$, where $h > 0$. Similarly, left hand side derivative of $|x|$ at zero is -1, that is, $\lim_{h \to 0-} \frac{|0+h|-|0|}{h} = -1$, where $h < 0$.

Remarks:

(i) A function has a derivative at a point x_0 if the slopes of the secant lines through $P(x_0, f(x_0))$ and a nearby point Q on the graph approach a limit as Q approaches P.

(ii) Whenever the secants fail to take up a limiting position or become vertical as Q approaches P, the derivative does not exist.

(iii) Differentiable function is always continuous.

(iv) A function is continuous at every point where it has a derivative.

(v) If f has derivative at $x = c$, then f is continuous at $x = c$, where converse is not true. That is, a function need not have derivative at a point where it is continuous for example $y = |x|$.

(vi) If a and b are any two points in an interval on which f is differentiable, then f' takes each value between $f'(a)$ and $f'(b)$.

(vii) The average rate of change of a function $f(x)$ with respect to x over the interval from x_0 to $x_0 + h$ is $\frac{f(x_0+h)-f(x_0)}{h}$.

(viii) The instantaneous rate of change of a function $f(x)$ with respect to x at x_0 is $f'(x_0) = \lim_{h \to 0} \frac{f(x_0+h)-f(x_0)}{h}$ provided the limit exists.

(ix) The sensitivity to change is when a small change in x results in a large change in the value of the function $f(x)$, and it means that the function is relatively sensitive to changes in x. The derivative function $f'(x)$ is a measure of the sensitivity to the change in x.

(x) Economists use the terms marginal for the rate of change and derivatives.

Example 20 *Let $f(x)$ be the cost to produce x units in a specified period. The cost of producing h more units is $f(x+h)$. Then, the average increase in costs of producing h more units is $\frac{f(x+h)-f(x)}{h}$. The limit of this as $h \to 0$ is the marginal cost of producing h units when the current production level is x. Where the cost of producing additional h units is $f(x+h) - f(x)$. If $h = 1$, then $\frac{\Delta f}{\Delta x} = \frac{f(x+1)-f(x)}{1} = \frac{df}{dx}$, which is approximately equal to $\frac{df}{dx}$ at x. It is observed that the slope of $f(x)$ does not change significantly near x, then difference value is close to its limit, that is $\frac{df}{dx}$, even if $\Delta x = 1$. Hence we can say that approximation gives the best results in the case of large values of x.*

1.19 Extreme Values of Functions

Let function f be continuous at every point of a closed interval if f has both an absolute maximum and a minimum value in the interval. Some cases are given for continuous functions with absolute maxima and minima in a closed interval [a,b].

1.20 Local and Absolute (Global) Extrema

Absolute Extreme Values: Let f be a function with domain D. Then f has an absolute maximum value on D at a point C, say, if $f(x) \leq f(C), \forall x$ in D. Whereas an absolute minimum value on D at C if $f(x) \geq f(C), \forall x$ in D.

Local Extreme Values: Let f be a function with domain D. Then f has a local maximum value at an interior point C. say, if $f(x) \leq f(C), \forall x$ in some open interval containing C. Whereas function f has a local minimum value at an interior point C of its domain D, if $f(x) \geq f(C), \forall x$ in open interval somewhere containing C.

Noted that the first derivative theorem for local extreme values helps us to determine the extreme value of the function f. The function f has a local minimum or maximum value at an interior point C of its domain D. That is, if f' is defined at C, then $f'(C) = 0$. Hence we can conclude that a function can have a local or global maximum or minimum value, where there are

(i) interior points where $f' = 0$

(ii) interior points where f' is undefined

(iii) endpoints of the domain D of f.

It means an interior point of the domain of function f where f' is zero or undefined is a critical point of f.

In summary, the only domain points where function f can assume extreme values are critical points and endpoints.

Example 21 *Let the function $f(x) = x^2$, be such that it is differentiable over its entire domain D. The critical point is where $f' = 2x = 0 \Rightarrow x = 0$. Therefore, the critical point value $f(0) = 0$, and end point values $f(-2) = f(2) = 4$, $f(1) = 1$. Hence, the function has an absolute maximum value 4 at $x = 2$, and an absolute minimum value 0 at $x = 0$.*

The function $f(x) = x^3$ has no extremum at $x = 0$, while $f' = 3x^2 = 0$ at $x = 0$. The function $f(x) = x^{1/3}$ has no extremum at $x = 0$, while $f' = \frac{1}{3}x^{-2/3}$ is undefined at $x = 0$. It means that every time critical points of the functions do not give an extreme value for the function f over the domain D.

Theorem 1.1

Rolle's Theorem

This theorem tells us that a differentiable curve has at least one horizontal tangent between two points where it crosses the x-axis. It may be just one, or it may be more. According to theorem, suppose that $y = f(x)$ is continuous at every point of the closed interval [a,b] and differentiable at every point of its interior (a,b). If $f(a) = f(b) = 0$, then there is at least one value c in (a,b) at which $f'(c)$ is zero. If both maximum and minimum are at a or b, f is constant, f' is zero and c can be taken anywhere in the interval.

Example 22 *Let $f(x) = \frac{x^3}{3} - 3x$, the function is continuous at every point [-3,3] and is differentiable at every point of (-3,3). Since $f(3) = f(-3) = 0$. By applying the theorem, $f'(x) = 0 = x^2 - 3 \Rightarrow x = \sqrt{3}$ and $x = -\sqrt{3}$. It means function $f'(x)$ is zero twice in the interval (-3,3), once at $x = \sqrt{3}$ and $x = -\sqrt{3}$. Hence the curve has horizontal tangents between the points where it crosses the x-axis.*

Theorem 1.2

Mean Value Theorem

According to the theorem, there is a point somewhere between A and B, where the curve has at least one tangent parallel to the chord AB. Lets suppose $y = f(x)$ is continuous in a closed interval [a,b] and differentiable in the interior on the interval (a,b). Then, there will be at least one point c in (a,b) at which $f'(c) = \frac{f(b)-f(a)}{b-a}$. If $\frac{f(b)-f(a)}{b-a}$ is the average change in f over [a,b] and $f'(c)$ is an instantaneous change, then according to the mean value theorem, there is some interior point where the instantaneous change must be equal to the average change over the entire interval.

Example 23 Let $f(x) = x^2$ be continuous for $0 \leq x \leq 2$ and differentiable for $0 < x < 2$. Since $f(0) = 0$ and $f(2) = 4$. According to mean value theorem there is a point c in the interval where derivative $f'(x) = 2x$ must have the value $\frac{4-0}{2-0} = 2$. Hence $c = 1$.

Let the another function $f(x) = \sqrt{1-x^2}$ satisfy the assumption of the mean value theorem in $[-1,1]$ even though f is not differentiable at -1 and 1.

1.21 Increasing Functions and Decreasing Functions

Let f be a function defined on an interval and let a and b be any two points in the interval. If f increases in the interval it means $a < b \Rightarrow f(a) < f(b)$. If f decreases in the interval it means $a < b \Rightarrow f(a) > f(b)$.

The first derivative test is also used for analysis of the function behaviour, that is, the function is increasing or decreasing. Suppose that f is continuous in [a,b] and differentiable on (a,b). If $f' > 0$ at each point of (a,b), then f increases on [a,b]. If $f' < 0$ at each point of (a,b), then f decreases on [a,b].

Some examples for increasing, decreasing, or stationary functions are:

Example 24 Let $f(x) = x^2$ decrease in $(-\infty, 0)$ where $f'(x) = 2x < 0$. The function increases in $(0, \infty)$ where $f'(x) = 2x > 0$.

Example 25 Let $f(x) = 3x^2 - 14x + 5$. Check the function at $x = 4$.

$$f'(x) = 6x - 14$$
$$f'(4) = 10 \geq 0,$$

This shows that the function is increasing at $x = 4$.

Example 26 Let $f(x) = x^3 - 7x^2 + 6x - 2$. Check the function at $x = 4$.

$$f'(x) = 3x^2 - 14x + 6$$
$$f'(4) = -2 \leq 0,$$

This shows that the function is decreasing at $x = 4$.

Example 27 Let $f(x) = x^4 - 6x^3 + 4x^2 - 13$. Check the function at $x = 4$.

$$f'(x) = 4x^3 - 18x^2 + 8x$$
$$f'(4) = 0,$$

This shows that the function is stationary at $x = 4$.

Example 28 *Let* $f(x) = -2x^3 + 4x^2 + 9x - 15$. *Check the function at* $x = 3$ *is convex or concave.*

$$f'(x) = -6x^2 + 8x + 9$$
$$f''(x) = -12x + 8$$
$$f''(3) = -28 \leq 0,$$

This shows that the function is concave.

Example 29 *Let* $f(x) = (5x^2 - 8)^2$. *Check the function at* $x = 3$ *is convex or concave.*

$$f'(x) = 100x^3 - 160x$$
$$f''(x) = 300x^2 - 160$$
$$f''(3) = 2540 \geq 0,$$

This shows that the function is convex.

1.22 Tests Based on First Derivative

The following test applies to a continuous function to search a local extreme value.

1. At a critical point: We have the three following situations

    (i) If f' changes from positive to negative at c, that is, $f' > 0$ for $x < c$ and $f' < 0$ for $x > c$, then f has a local maximum value at c (see Figs. (1.8) & (1.9)).

    (ii) If f' changes from negative to positive at c, that is, $f' < 0$ for $x < c$ and $f' > 0$ for $x > c$, then f has a local minimum value at c (see Figs. (1.10) & (1.11)).

    (iii) If f' does not change sign at c, that is, f' has the same sign on both sides of c, then f has no local extreme value at c (see Figs. (1.12) & (1.13)).

2. At a left end point: If $f' < 0$ for $x > c$, then f has a local maximum value at c (see Fig. (1.14)). If $f' > 0$ for $x > c$, then f has a local minimum value at c (see Fig. (1.15)).

3. At a right end point: If $f' < 0$ for $x > c$, then f has a local minimum value at c (see Fig. (1.17)). If $f' > 0$ for $x > c$, then f has a local maximum value at c (see Fig. (1.18)).

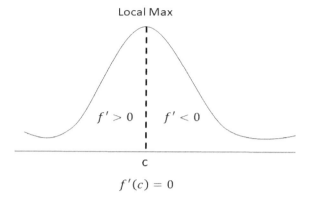

Figure 1.8: Local Maxima.

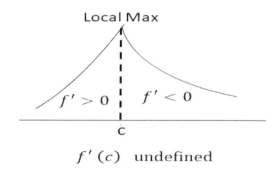

Figure 1.9: Local Maxima.

1.23 Concavity

The graph of a differentiable function $f(x)$ is concave up on an interval where f' is an increasing function and concave down on an interval where f' is a decreasing function.

According to the second derivative test for concavity, if the function $f(x)$ has a second derivative, then we can conclude that f' is a increasing function if $f'' > 0$ and it is decreases if $f'' < 0$.

Example 30 *Let $f(x) = x^3$ is concave down in $(-\infty, 0)$ where $f''(x) = 6x < 0$ and concave up in $(0, \infty)$ where $f''(x) = 6x > 0$.*

Example 31 *Let $f(x) = x^2$ is concave up in $(-\infty, 0)$ where $f''(x) = 2$ and concave up in $(0, \infty)$ where $f''(x) = 2$. It means the function is concave up on every point of interval.*

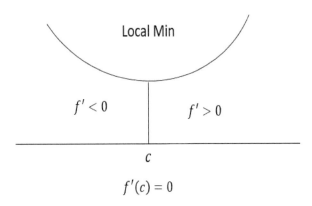

Figure 1.10: Local Minima.

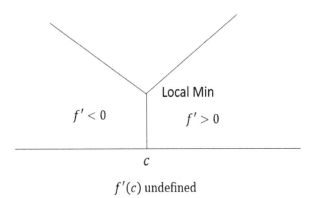

Figure 1.11: Local Minima.

1.24 Inflection Point

A point where the function graph has a tangent line and where the concavity changes are known as a point of inflection. In other words, a point of inflection on a curve is a point where $f'' > 0$ on one side and $f'' < 0$ on the other side. At such a point, f'' is either zero or undefined.

Example 32 *Let $f(x) = x^4$ has no inflection point at $x = 0$. Even though $f''(x) = 12x^2$ is zero and it does not change sign. Hence, the function has no inflection point at the origin even though $f''(x) = 0$.*

Example 33 *Let function $f(x) = 2x^4 - 16x^3 + 32x^2 + 15$. Check for relative maxima or minima.*

$$f'(x) = 8x^3 - 48x^2 + 64x = 0$$
$$= 8x(x-2)(x-4) = 0$$

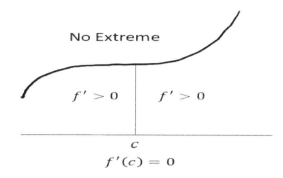

Figure 1.12: No Extreme Point.

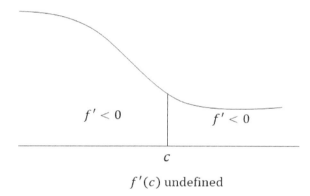

$f'(c)$ undefined

Figure 1.13: Undefined.

or $x = 0, x = 2$, and $x = 4$ are the critical values for the function $f(x)$.

$$f''(x) = 24x^2 - 96x + 64$$

$f''(0) = 64 \geq 0$ convex, relative minimum

$f''(2) = -32 \leq 0$ concave, relative maximum

$f''(4) = 64 \geq 0$ convex, relative minimum

Example 34 *Let function* $f(x) = -(x-8)^4$. *Check for relative maxima or minima.*

$$f'(x) = -4(x-8)^3 = 0$$

or $x = 8$ is the critical value for the function $f(x)$.

$$f''(x) = -12(x-8)^2$$

$$f''(8) = 0$$

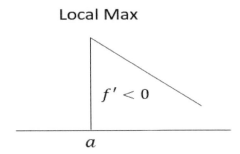

Figure 1.14: Local Maxima.

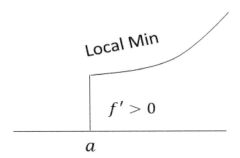

Figure 1.15: Local Minima.

test inconclusive, continue to take successively higher derivatives.

$$f'''(x) = -24(x-8)$$

$f'''(8) = 0$ test inconclusive

$$f''''(x) = -24$$

$f''''(8) = -24 \leq 0$, concave relative maximum

1.25 Linearization Approximation

The approximation of a function is called linearization, and it is based on tangent lines. In some cases, we need to approximate the complicated function with simpler ones. In the Fig. (1.16), the tangent line passes through the point $(a, f(a))$, so its point-slope equation is $y = f(a) + f'(a)(x-a)$. It is also called as the tangent, since tangent is the graph of the function.

In another words, we can say that if f is differentiable at $x = a$, then the approximation function $L(x) = f(a) + f'(a)(x-a)$ is the linearization of f at a.

Basic Calculus ■ 21

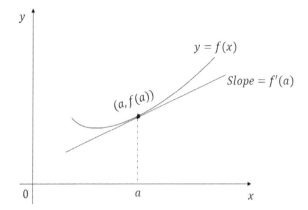

Figure 1.16: Linearization Approximation Tangent Line.

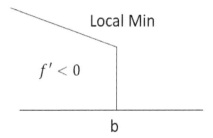

Figure 1.17: Local Minima.

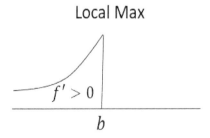

Figure 1.18: Local Maxima.

Example 35 Let $f(x) = \sqrt{1+x}$ at $x = 0$, then $f'(x) = \frac{1}{2}(1+x)^{-1/2}$. We have $f(0) = 1, f'(0) = 1/2$. Hence $L(x) = f(a) + f'(a)(x-a) = 1 + \frac{x}{2}$.
 Let $f(x) = \sqrt{1+x}$ at $x = 3$, we have $f(3) = 2$, $f'(3) = 1/4$. Hence $L(x) = f(a) + f'(a)(x-a) = \frac{x+5}{4}$.

Chapter 2

Matrix and Determinant Algebra

In this chapter, the preliminaries of a matrix will be discussed. Matrix algebra is very useful in the optimization techniques.

2.1 Matrix

A matrix is a rectangular arrangement of elements (real or complex numbers), arranged in m rows and n columns. That is

$$A = \begin{pmatrix} a_{11} & a_{12} & \cdots & a_{1j} & \cdots & a_{1n} \\ a_{21} & a_{22} & \cdots & a_{2j} & \cdots & a_{2n} \\ \vdots & \vdots & & \vdots & & \vdots \\ a_{i1} & a_{i2} & & a_{ij} & & a_{in} \\ \vdots & \vdots & & \vdots & & \vdots \\ a_{m1} & a_{m2} & \cdots & a_{mj} & \cdots & a_{mn} \end{pmatrix} \quad (2.1)$$

Matrix A defined in Eqn. (2.1) is an $m \times n$ matrix with elements a_{ij} ($i = 1, 2, \ldots, m$; $j = 1, 2, \ldots, n$), where m denotes the number of rows and n denotes the number of columns of a matrix. If m=n then we say that matrix A defined in Eqn. (2.1) is a square matrix.

2.2 Types of Matrix

Null Matrix A matrix $A_{m \times n}$ is said to be a null matrix if all its elements are zero.

Diagonal Matrix A matrix $A_{m \times n}$ is said to be diagonal matrix if all the off-diagonal elements are zero and at least one diagonal element is non-zero.

Identity Matrix A square matrix $A_{m \times n}$ having all its diagonal elements equal to unity and all off-diagonal elements as zero.

Symmetric Matrix A matrix A is called symmetric if $A = A'$, that is, $a_{ij} = a_{ji}$. The definition implies that A is a square matrix.

Skew Symmetric Matrix A matrix A is called skew-symmetric if $A = -A'$, that is, $a_{ij} = -a_{ji}$. It is also a square matrix whose diagonal elements are zero.

2.3 Algebra of Matrix

This section explain some matrix algebra, laws, ranks, types and its properties.

2.3.1 Addition and Subtraction of Matrix

The sum of a matrix $A_{m \times n} = [a_{ij}]$ and a matrix $B_{m \times n} = [b_{ij}]$ is defined only when A and B have same number of rows and columns. The sum is then

$$A + B = [a_{ij}] + [b_{ij}] = C(say)$$

$A - B$ is defined as $A + (-B)$. It is called the difference of A and B.

2.3.2 Multiplication by a Scalar

When k is a real number, and A is a matrix, kA is defined to be the matrix, each element of which is k times the corresponding element of A. That is, $kA_{m \times n} = [ka_{ij}]$. In particular, if $k = -1$, then $[-1a_{ij}] = -A$.

2.3.3 Matrix Multiplication

The product $[A_{s \times r}][B_{r \times t}]$ is defined only when the number of columns in A is equal to the number of rows in B, and the product is a matrix whose element in the $(i, j)^{th}$ position is the inner product of the ith row of A by the jth column of B. The resulting matrix is $s \times t$. Obviously, $AB \neq BA$.
If $A = [a_{ij}], B = [b_{ij}]$, then $AB = C_{s \times t} = (c_{ij})$, where

$$c_{ij} = \sum_{k=1}^{r} a_{ik} b_{kj}, \quad i = 1, 2, \ldots, s; \quad j = 1, 2, \ldots, t$$

2.3.4 Laws of Algebra

Addition:

(i) Associative law holds
$$(A+B)+C = A+(B+C)$$

(ii) Commutative law holds
$$A+B = B+A$$

(iii) Distributive law holds
$$k(A+B) = kA+kB$$
$$(c+d)A = cA+dA$$
$$-(A-B) = -A+B$$

Multiplication:

(i) Associative law holds
$$A(BC) = (AB)C$$

(ii) Distributive law holds
$$A(B+C) = AB+AC$$

(iii) Commutative law does not holds
$$AB \neq BA$$

2.4 Transpose of Matrix

The transpose of a matrix $A = [a_{ij}]$ of order $m \times n$ denoted by $A' = [a'_{ij}]$ is a matrix of order $n \times m$ such that $a'_{ij} = a_{ji}$ that is $(i,j)^{th}$ element of A' is the $(j,i)^{th}$ element of A, or equivalently, the rows of A' are the columns of A.

Propositions

(i) $(A+B)' = A'+B'$

(ii) $(AB)' = B' \times A'$

(iii) $(A_1, A_2, \ldots, A_n)' = A'_n, \ldots, A'_2, A'_1$

(iv) $(kA)' = kA'$

(v) $(A')' = A$

2.5 Trace of Matrix

The trace of a square matrix A, denoted by trA, is defined as the sum of its diagonal elements, that is, $trA = \sum_i a_{ii}$.
Proposition:

(i) If A and B are square matrices of the same order, then $tr(A+B) = trA + trB$.

$$tr(A+B) = \sum_i (a_{ii} + b_{ii}) = \sum_i a_{ii} + \sum_i b_{ii} = trA + trB$$

(ii) If C and D are such that CD and DC are both defined, then CD and DC are both squares and $trCD = trDC$. Let C be $m \times n$ and D be $n \times m$. Then CD is $m \times m$ and DC is $n \times n$. Then

$$trCD = \sum_{i=1}^{m} \left(\sum_{k=1}^{n} c_{ik} d_{ki} \right)$$

$$trDC = \sum_{k=1}^{n} \left(\sum_{i=1}^{m} d_{ki} c_{ik} \right)$$

2.6 Non-Singular Matrix

A matrix A is called non-singular if there exists a matrix B such that $AB = BA = I$. This definition implies that both A and B must be square matrices of the same order.

2.7 Inverse of a Matrix

Given a matrix A, if there exists a square matrix B such that $AB = BA = I$, then B is called the inverse of A.
Properties:

(i) Inverse matrix is unique
Let B and C be two inverses of A. Then

$$AB = BA = I = AC = CA$$

Now,
$$CAB = (CA)B = IB = B$$

and
$$C(AB) = CI = C$$

$$\Rightarrow B = C$$

(ii) $(AB)^{-1} = B^{-1}A^{-1}$ since $B^{-1}A^{-1}AB = B^{-1}IB = B^{-1}B = I$

(iii) $(A^{-1})' = (A')^{-1}$

2.8 Orthogonal Matrix

A square matrix C is said to be an orthogonal matrix if

$$C'C = CC' = I$$

Properties:

(i) If A be any matrix, then AA' and $A'A$ are symmetric.

(ii) If A and B are symmetric, then $A + B$ is also symmetric.

(iii) If A and B are skew symmetric, then $A + B$ is also skew symmetric.

(iv) If A and B are symmetric matrices, AB is symmetric $\iff$ A and B commute.

(v) If A and B are n-rowed orthogonal matrices, then AB and BA are also orthogonal.

(vi) If A is orthogonal matrix, then A' is also orthogonal.

2.9 Rank of Matrix

The rank of an $m \times n$ matrix A, denoted by $R(A)$, is the maximum number of independent columns in A (or more precisely the column rank).

Theorem 2.1
If A and B are matrices such that the product AB is defined, then $R(AB) \leq \min\{R(A), R(B)\}$.

Theorem 2.2
The row rank of a matrix is equal to its column rank.

2.10 Determinants and Non-Singularity

A determinant $|A|$ of a 2×2 matrix, called a second-order determinant. Let a general 2×2 matrix as:

$$A = \begin{bmatrix} a_{11} & a_{12} \\ a_{21} & a_{22} \end{bmatrix}$$

The determinant is $|A| = \begin{vmatrix} a_{11} & a_{12} \\ a_{21} & a_{22} \end{vmatrix} = a_{11}a_{22} - a_{12}a_{21}$

The determinant is a single number or scalar value of a square matrix. If the determinant of a matrix, $|A| = 0$, then the matrix is said to be a singular matrix.

- A singular matrix is one in which there exists linear dependence between at least two rows or columns.

- If $|A| \neq 0$, the matrix is non-singular and all its rows and columns are linearly independent.

- If linearly dependence exists in a system of equations, then the system as a whole will have an infinite number of possible solutions, making a unique solution impossible.

Results:

1. If $|A| = 0$, the matrix is singular and there is linear dependence among the equations. No unique solution is possible.

2. If $|A| \neq 0$, the matrix is non-singular and there is no linear dependence among the equations. A unique solution can be found.

3. If $R(A) = n$ of a square matrix $A_{n \times n}$ of order n, then it is a non-singular matrix and there is no linear dependence.

4. If $R(A) < n$, $A_{n \times n}$ is singular and there exist linear dependence.

Example 36 *If* $A = \begin{bmatrix} 6 & 4 \\ 7 & 9 \end{bmatrix}$, $B = \begin{bmatrix} 4 & 6 \\ 6 & 9 \end{bmatrix}$, *determine which matrix is singular or non-singular?*

Solution
$$|A| = 6 \times 9 - 7 \times 4 = 26$$

Since $|A| \neq 0$, the matrix is non-singular. That is, there is no linear dependence between any of its rows or columns. The rank of A, i.e., $R(A) = 2$.

Similarly,
$$|B| = 4 \times 9 - 6 \times 6 = 0$$

Since $|B| = 0$, it is a singular matrix, and linear dependence exists between rows and columns. It is further observed that row 2 and column 2 are equal to 1.5 times to row 1 and column 1, respectively. Hence $R(B) = 1$. ∎

2.11 Three Order Determinants

Consider a matrix of order 3×3

$$A = \begin{bmatrix} a_{11} & a_{12} & a_{13} \\ a_{21} & a_{22} & a_{23} \\ a_{31} & a_{32} & a_{33} \end{bmatrix}$$

The calculations for the determinant can be done about any one row or column of the matrix as follows:

$$|A| = a_{11} \begin{vmatrix} a_{22} & a_{23} \\ a_{32} & a_{33} \end{vmatrix} - a_{12} \begin{vmatrix} a_{21} & a_{23} \\ a_{31} & a_{33} \end{vmatrix} + a_{13} \begin{vmatrix} a_{21} & a_{22} \\ a_{31} & a_{32} \end{vmatrix}$$

$$= a_{11}(a_{22}a_{33} - a_{23}a_{32}) - a_{12}(a_{21}a_{33} - a_{23}a_{31}) + a_{13}(a_{21}a_{32} - a_{22}a_{31})$$

$$= a, \ scalar \ value$$

where the determinant of the sub-matrix formed by deleting the ith row and j^{th} column of the matrix is known as minor denoted as $|M_{ij}|$. That is,

$$|M_{11}| = \begin{vmatrix} a_{22} & a_{23} \\ a_{32} & a_{33} \end{vmatrix}, \ |M_{12}| = \begin{vmatrix} a_{21} & a_{23} \\ a_{31} & a_{33} \end{vmatrix}, \ |M_{13}| = \begin{vmatrix} a_{21} & a_{22} \\ a_{31} & a_{32} \end{vmatrix}$$

Therefore,

$$|A| = a_{11}|M_{11}| - a_{12}|M_{12}| + a_{13}|M_{13}|$$

The rule for the sign of a cofactor is

$$|c_{ij}| = (-1)^{i+j}|M_{ij}|$$

- If $i+j$=even $\Rightarrow |c_{ij}| = |M_{ij}|$
- If $i+j$=odd $\Rightarrow |c_{ij}| = -|M_{ij}|$

Properties:

1. Adding or subtracting any non zero multiples of one row (or column) from another row (or column) will not affect the determinant.

2. Interchanging, any two rows or columns of a matrix, will change the sign, but not the determinant's absolute value.

3. Multiplying the elements of any row or column by a constant will cause the determinant to be multiplied by the constant.

4. The determinant of a triangular matrix, that is, a matrix with zero elements everywhere above or below the principal diagonal is equal to the product of the elements on the principal diagonal.

Example 37 *Consider a lower triangular matrix* **A**, *which has zero elements below the principal diagonal. That is,*

$$\mathbf{A} = \begin{bmatrix} 2 & -5 & -1 \\ 0 & 3 & 6 \\ 0 & 0 & -7 \end{bmatrix} = |A| = 2 \times 3 \times (-7) = -42$$

or $|A| = 2(-21 - 0) - (-5)(0 - 0) - 1(0 - 0) = -42$

5. If all the elements of any row or column are zero, the determinant is zero.

Example 38 *Consider matrix* **A** *given by*

$$\mathbf{A} = \begin{bmatrix} 12 & 16 & 13 \\ 0 & 0 & 0 \\ -15 & 20 & -9 \end{bmatrix} = 12(0 - 0) - 16(0 - 0) + 13(0 - 0) = 0$$

Since all the elements of row 2 are zero, the matrix is actually a 2×3 matrix not a 3×3 matrix. As we know that only a square matrix have determinants.

6. The determinant of a matrix equals to the determinant of its transpose $|A| = |A'|$.

Example 39 *Let*

$$A = \begin{bmatrix} 2 & 5 & 1 \\ 3 & 2 & 4 \\ 1 & 4 & 2 \end{bmatrix}$$

$$|A| = -24$$

$$A' = \begin{bmatrix} 2 & 3 & 1 \\ 5 & 2 & 4 \\ 1 & 4 & 2 \end{bmatrix}$$

$$|A'| = -24$$

7. If two rows or columns are identical or proportional, that is, linearly dependent, the determinant is zero.

8. Inverse of a matrix is unique.

9. A square matrix A has an inverse $\iff |A| \neq 0$, that is, a non-singular matrix has an inverse.

10. If A is non-singular, then $(A^{-1})^{-1} = A$.

11. If A and B are two non-singular matrices of the same order, then $(AB)^{-1} = B^{-1}A^{-1}$.

12. If A is non-singular, then $(A')^{-1} = (A^{-1})'$.

13. If the product of two square matrices (i.e., AB) is zero, then either $A = 0$ or $B = 0$ or both A and B are singular matrices.

14. $Rank(A) = 0$, iff $A = 0$

15. $Rank(A) \geq 1$, iff $A \neq 0$

16. $Rank(A) \leq m$, if A is $m \times n$ matrix such that $m \leq n$.

17. $Rank(A) < n$, if A is $m \times n$ matrix such that $n \leq m$.

18. $Rank(A') = Rank(A)$, if A' is transpose of A.

19. Rank of a matrix is 1, if every entry of its element is unity.

   Example 40 Let
   $$A = \begin{bmatrix} 1 & 1 & 1 & 1 & 1 \\ 1 & 1 & 1 & 1 & 1 \\ 1 & 1 & 1 & 1 & 1 \end{bmatrix},$$
   hence $R(A)=1$.

20. Two equivalent matrices have the same rank.

21. The rank of a product of two matrices cannot exceed the rank of either matrix. That is,
$$Rank(AB) \leq Rank(A)$$
$$Rank(AB) \leq Rank(B)$$

2.12 Linear Equations

Consider a system of linear equations $\sum_{j=1}^{n} a_{ij}x_j = b_i$, $i = 1, 2, \ldots, m$ where $x'_j s$ are referred to as variables or unknowns, and the $a'_{ij}s$ and $b'_i s$ are constants. A solution to the linear system of m equations in n variables is a set of values for the unknowns that satisfies each of the system's m equations.

Example 41 Consider the system of linear equations
$$2x_1 + x_2 = 4$$
$$4x_1 - 2x_2 = 0$$

Show that $x_1 = 1, x_2 = 2$ is a solution to the linear system and that $x_1 = 3, x_2 = 2$ is not a solution to linear equations.

If $\underline{x} = (1,2)'$ is a solution, then it must satisfy both equations, that is, $2 \times 1 + 2 = 4$ and $4 \times 1 - 2 \times 2 = 0$. The vector $\underline{x} = (3,2)'$ fails to satisfy the equations, hence it is not the solution to the equations.

2.13 Vector Algebra

A vector is a quantity with a magnitude and direction in practice. A column vector can be converted into a row vector by a simple transpose or vice versa, that is,

$$\begin{pmatrix} x \\ y \\ z \end{pmatrix}^T = (x, y, z)$$

The magnitude or length of a three-dimensional vector in its cartesian form is $\sqrt{x^2 + y^2 + z^2}$.

In general, a vector in an n-dimensional space $(n \geq 1)$ is a row vector $\underline{x} = (x_1, x_2, \ldots, x_n)$. It's length can be defined as

$$||x|| = \sqrt{x_1^2 + x_2^2 + \cdots + x_n^2}$$

The addition or substraction of two vectors u and v are the addition or substraction of their corresponding components, that is,

$$u \pm v = \begin{pmatrix} u_1 \\ u_2 \\ \vdots \\ u_n \end{pmatrix} \pm \begin{pmatrix} v_1 \\ v_2 \\ \vdots \\ v_n \end{pmatrix} = \begin{pmatrix} u_1 \pm v_1 \\ u_2 \pm v_2 \\ \vdots \\ u_n \pm v_n \end{pmatrix}$$

The inner product of two vectors u and v is defined as $u^T v = u.v = \sum_{i=1}^{n} u_i v_i = u_1 v_1 + u_2 v_2 + \ldots + u_n v_n$.

2.13.1 Metric Spaces

Let X be a non empty set. A metric (or distance function) on X is a mapping $d : X \otimes x \to R$ which satisfies the following axioms $\forall\ x, y, z \in x$.

Axiom 1: $d(x,x) = 0$

It means that the distance of a point from itself is zero.

Axiom 2: $d(x,y) = 0 \Rightarrow x = y$

It means that if the distance is zero, the two points are the same.

Axiom 3: $d(x,y) = d(y,x)$

It states that the distance does not depend on the order of the points.

Axiom 4: $d(x,z) \leq d(x,y) + d(y,z)$

It is called the triangle inequality. Equality can be hold only when three points are in a straight line.

2.13.2 Normed Linear Spaces

A linear space together with a norm is called a normed linear space. It is a special metric space with high interaction of the algebraic and geometric structures. It is essential that the metric (or distance) function respect the linearity of the space.

For any linear space X, a norm $\| x \|$ is a measure of the size of the elements satisfying the following conditions

1. $\| X \| \geq 0$
2. $\| X \| = 0$ iff $X = 0$
3. $\| \alpha X \| = |\alpha| \| X \| \quad \forall \ \alpha \in R$
4. $\| X + Y \| \leq \| X \| + \| Y \|$ triangle inequality

A norm on a linear space X induces a metric on X in which the distance between any two elements (points) is given by the norm of their difference

$$d(x,y) = \| X - Y \|$$

Notes:

1. Every inner product defines a norm given by $\| X \| = \sqrt{X'X}$, consequently every inner product space is a normed linear space.
2. A finite-dimensional inner product space is called a Euclidean space (E^n).
3. A complete inner product space is called a Hilbert space.
4. R^n is a Euclidean space, with inner product $X'Y = \sum_{i=1}^{n} X_i Y_i$.
5. Every Euclidean space is complete that is, a Hilbert space.
6. The function $\| X \| = \sqrt{X'X}$ is a norm on X.
7. For every X, Y in an inner product space, $|X'Y| \leq \| X \| \| Y \|$ (Cauchy-Schwartz Inequality).
8. (Existence of extreme points) A non empty compact convex set in an inner product space has at least one extreme point.

2.13.3 Norms

For an n dimensional vector x, we can define a p norm or L_p-norm as

$$||x||_p = (|x_1|^p + |x_2|^p + \cdots + |x_n|^p)^{1/p} = \left(\sum_{i=1}^{n} |x_i|^p\right)^{1/p}, \; p > 0$$

For p = 1,
$$||x||_1 = |x_1|^1 + |x_2|^1 + \cdots + |x_n|^1 = \sum_{i=1}^{n} |x_i|$$

For p = 2,
$$||x||_2 = (|x_1|^2 + |x_2|^2 + \cdots + |x_n|^2)^{1/2} = \sqrt{\sum_{i=1}^{n} |x_i|^2}$$

For $p = \infty$, it becomes
$$||x||_\infty = max\{|x_1|, |x_2|, \cdots, |x_n|\} = x_{max}$$

In general, for any two vectors u and v in the same space, we have the following equality
$$||u||_p + ||v||_p \geq ||u+v||_p, \; (p \geq 0)$$

Example 42 *For two vectors* $u = [1,2,3]^T$ *and* $v = [1,-2,-2]$, *we have*

$$u^T v = 1 \times 1 + 2 \times (-2) + 3 \times (-2) = -10$$

$$||u||_1 = |1| + |2| + |3| = 6, \; ||v||_1 = |1| + |-2| + |-2| = 5$$
$$||u||_2 = \sqrt{1^2 + 2^2 + 3^2} = \sqrt{14}, \; ||v||_2 = \sqrt{1^2 + (-2)^2 + (-2)^2} = \sqrt{9} = 3$$
$$||u||_\infty = max\{|1|,|2|,|3|\} = 3, \; ||v||_\infty = max\{|1|,|-2|,|-2|\} = 2$$

and
$$w = u + v = [1+1, 2+(-2), 3+(-2)]^T = [2,0,1]^T$$

whose norms are
$$||w||_1 = 3, \; ||w||_\infty = max\{|2|,|0|,|1|\} = 2,$$
$$||w||_2 = \sqrt{2^2 + 0^2 + 1^2} = \sqrt{5}$$

2.13.4 Forbenlus Norm

Similar to the norms for vectors, we can define norms for matrix named as Forbenlus Norm. Let matrix of size $m \times n$, be the Forbenlus norm is defined as $||A||_F = \sqrt{\sum_{i=1}^{m} \sum_{j=1}^{n} |a_{ij}|^2}$, which is equivalent to $\sqrt{tr(A^T A)} = \sqrt{diag(A^T A)}$.

The maximum absolute column sum norm is defined as

$$||A||_1 = \max_{1 \leq j \leq n} \sum_{i=1}^{m} |a_{ij}|$$

Similarly, the maximum absolute row sum norm is defined as

$$||A||_\infty = \max_{1 \leq i \leq m} \sum_{j=1}^{n} |a_{ij}|$$

Example 43 *Let* $A = \begin{pmatrix} 2 & -2 & 4 \\ -6 & 1 & 7 \end{pmatrix}$, *we have*

$$||A||_1 = \max\{|2|+|-6|, |-2|+|1|, |4|+|7|\} = \max\{8, 3, 11\} = 11,$$

$$||A||_\infty = \max\{|2|+|-2|+|4|, |-6|+|1|+|7|\} = \max\{10, 14\} = 14,$$

and

$$||A||_F = \sqrt{|1|^2 + |-2|^2 + |4|^2 + |-6|^2 + |1|^2 + |7|^2} = \sqrt{107}$$

2.13.5 Eigenvalues Vectors

An eigenvalue, say λ of a square matrix is defined as $Au = \lambda u$, where a non zero eigenvector u exists for a corresponding λ. An $A_{n \times n}$ matrix can have at most n different eigenvalues and thus n corresponding eigenvectors. To obtain a row trial solution $u \neq 0$, the matrix $A - \lambda I$ is required not to be invertible, where I is an identity $n \times n$ matrix. That is $(A - \lambda I)u = 0$ if $u \neq 0$, then $det(A - \lambda I) = 0$, which is equivalent to a polynomial of order n, and it is called characteristic polynomial. All the eigenvalues form a set which is known as the spectrum of matrix A.

Example 44 *Let* $A = \begin{pmatrix} 2 & 3 \\ 3 & 4 \end{pmatrix}$, *its eigenvalues are defined by*

$$det(A - \lambda I) = det \begin{vmatrix} 2-\lambda & 3 \\ 3 & 4-\lambda \end{vmatrix} = 0$$

which gives

$$(2-\lambda)(4-\lambda) - 9 = 0$$

or $\lambda^2 - 6\lambda - 1 = 0$

After solving the equation, the two eigenvalues are $\lambda_1 = 3 + \sqrt{10}$, $\lambda_2 = 3 - \sqrt{10}$. The eigenvector $u = (a \ b)^T$ corresponding to λ_1 is obtained by the definition $Au = \lambda_1 u$, that is,

$$\begin{pmatrix} 2 & 3 \\ 3 & 4 \end{pmatrix} \begin{pmatrix} a \\ b \end{pmatrix} = 3 + \sqrt{10} \begin{pmatrix} a \\ b \end{pmatrix}$$

This is equivalent to two equations, that is

$$\left. \begin{array}{r} 2a + 3b = (3 + \sqrt{10})a \\ 3a + 4b = (3 + \sqrt{10})b \end{array} \right\} \quad (2.2)$$

Similarly, the eigenvector $u_2 = (c \ d)^T$ for λ_2 can be obtained as:

$$\begin{pmatrix} 2 & 3 \\ 3 & 4 \end{pmatrix} \begin{pmatrix} c \\ d \end{pmatrix} = 3 - \sqrt{10} \begin{pmatrix} c \\ d \end{pmatrix}$$

The equivalent equations are:

$$\left. \begin{array}{r} 2c + 3d = (3 - \sqrt{10})c \\ 3c + 4d = (3 - \sqrt{10})d \end{array} \right\} \quad (2.3)$$

After solving Eqns. (2.2) & (2.3), the corresponding eigenvectors are:

$$u_1 = \begin{pmatrix} \frac{(1 + \sqrt{10})}{3} \end{pmatrix}, \quad u_2 = \begin{pmatrix} \frac{(1 - \sqrt{10})}{3} \end{pmatrix}.$$

Chapter 3
LINGO-18: Optimization Software

3.1 Introduction

The linear, non-linear, quadratic, stochastic, and integer optimization problems can be solved using a simplified tool named LINGO. LINGO not only solves the problem but also analyzes them. LINGO is a user-friendly software; even a beginner can write programs and solve them. LINGO gives a completely unified package that covers a powerful language for demonstrating optimization models, a full-featured environment for building and editing problems, and a set of fast built-in solvers.

The latest version of the software available is LINGO 18.0 for Windows (32/64 bit size), Mac (64-bit size), and Linux (64-bit size) released in 2018.

3.1.1 Key Benefits of LINGO

Some key benefits of using LINGO software are:

1. Easy model expression.

2. Convenient data options.

3. Powerful solvers.

4. Interactive modeling and creation of turn-key applications.

5. Detailed documentation and help.

3.1.2 Installing LINGO

To install the LINGO software one can use any option given below:

Option 1 Put CD into drive, run setup, and follow the instructions.

Option 2 Directly download the free version of LINGO from the given link below https://www.lindo.com/lindoforms/downlingo.html.

3.1.3 Window Types in LINGO

After installation, the icon of LINGO will appear on the desktop. Some kinds of windows and toolbar of LINGO are as follows:

- LINGO Model-LINGO1: A blank page for writing the LINGO codes as shown in Fig. (3.1).

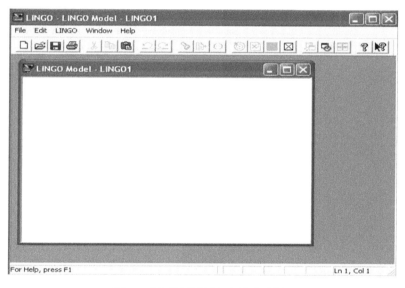

Figure 3.1: LINGO Model-LINGO1.

- Syntax Error Window: If any error occurred while typing the problem then the window shown in Fig. (3.2) appears.

- Solver status window: This window gives the details of the solution as shown in Fig. (3.3).

- Tool Bar: The Fig. (3.4) is the toolbar of LINGO in which all the icons are displayed.

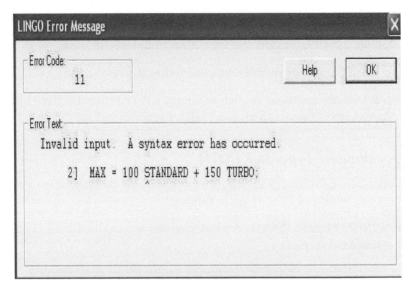

Figure 3.2: Syntax Error Window.

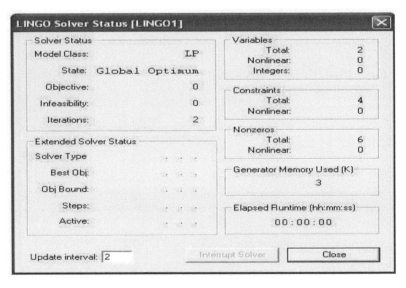

Figure 3.3: Solver Status Window.

3.1.4 Model Creation and Solution in LINGO

Two types of models can be created in LINGO, one is the direct model, and the other is the model using sets. The steps of creation and solution of the problem through the direct model is as follows:

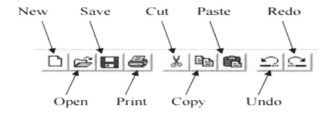

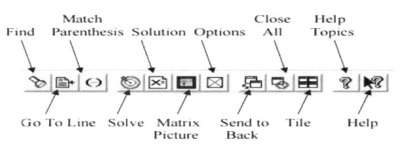

Figure 3.4: Tool Bar.

Step 1 To open the window, double click on the LINGO icon. A blank window with the title "LINGO Model-LINGO1" will appear. Based on requirements, a new window can be opened by selecting new from the file menu.

Step 2 In LINGO syntax (screen shown in Fig. (3.1)) enter the problem as follows:
Model:
!objective function;
Min=2 * x1 + x2;
!constraints;
3 * x1 + x2 =3;
4 * x1 + 3 * x2 >= 6;
x1 + 2 * x2 <= 3;
End

Step 3 From the menu, select the 'Solve' command or directly click on the solve icon shown on the toolbar to solve the problem.

Step 4 After completing the solution, a display, as shown in Fig. (3.3), will appear that displays the status of the solve command.

Step 5 Close the LINGO Solver Status window.

Step 6 Data overlying a display named "Solution Report-LINGO1" can be seen in which objective value, variable value, reduced cost, slack or surplus values, and dual prices are given.

Sets establish LINGO's modelling language—the essential structure block of the program's most remarkable capacities. With a comprehension of sets, you can compose a progression of comparable imperatives in a solitary proclamation and express long, complex equations briefly. Sets permit you to communicate your biggest models rapidly and without any problem. In bigger models, you'll experience the need to communicate a gathering of a few fundamentally same calculations or constraints. Luckily, LINGO's capacity to deal with sets of data permits you to perform such tasks productively. The sets section begins with the keyword 'SETS:' (including the colon) and ends with the keyword 'ENDSETS' model called the sets section. Specify the following to define a set in a sets section.

- set name,
- its members, i.e., the objects contained in the set (optional), and
- any attributes the members of the set may have (optional).

Now, the overall syntax for the SETS are:

$$setname[/member\_list/][: attribute\_list];$$

$$@function(setname[(set\_index\_list)[|conditional\_qualifier]] : expression\_list);$$

where @function-corresponds to any one of the four-set looping functions listed below:

- setname- the name of the set you want to loop over.
- set_index_list- optional, which is used to create a list of indices.
- setname-name chosen to designate the set.
- $member\_list$- list of the members that constitute the set.

An example to demonstrate the LINGO model construction through SETS is as follows:

Example 45 *A manufacturing industry has three types of raw materials and four warehouses. Transportation cost per ton, capacities, and requirements are given in Table (3.1). Determine the optimum allocation of the raw materials to minimize the transportation cost.*

To derive the optimal transportation cost, the given transportation can be formulated as a LPP. Then, the formulated LPP is solved through LINGO software. The transportation problem is coded as follows in LINGO software.

Table 3.1: Data of Transportation Cost, Capacity and Demand.

		Warehouse				
		W_1	W_2	W_3	W_4	Capacity (tons)
	R_1	5	7	3	8	300
Raw Material	R_2	4	6	9	5	500
	R_3	2	6	4	5	200
	Demand	200	300	400	100	

LINGO Codes

MODEL:
!A 3 raw materials and 4 warehouses transportation problem;
SETS:
RAWMATERIAL/R1,R2,R3/:CAPACITY;
WAREHOUSE/W1,W2,W3,W4/:DEMAND;
LINKS(RAWMATERIAL,WAREHOUSE):COST, SHIP;
ENDSETS
!The objective function of minimizing the transportation cost;
MIN=@SUM(LINKS:COST(I,J)*SHIP(I,J));
!Demand constraints;
@FOR(WAREHOUSE(J):
@SUM(RAWMATERIAL(I):SHIP(I,J))>DEMAND(J));
!Capacity constraints;
@FOR(RAWMATERIAL(I):
@SUM(WAREHOUSE(J):SHIP(I,J))<CAPACITY(I));
!Here data is entered;
DATA:
CAPACITY=300,500,200;
DEMAND=200,300,400,100;
COST=5,7,3,8
4,6,9,5,
2,6,4,5
ENDDATA
END

Notes:

- Exclamation mark (!) is used to write the comments and these comments appear in green text.

- In blue text, specified LINGO operators appear, such as MODEL, SETS, MIN, MAX and so on.

- The rest of the text appears in black.

- Semi-colon (;) is used to end each LINGO statement.

- Names of the variables must start with a letter (A-Z) as they are not case sensitive. Other characters in the variable name may be letters, numbers (0-9), or the underscore character (_). Upto 32 characters in length can be used to name variables.

- To save the input data for further use, go to the File menu and select save.

- To open a new blank file, either go to the File menu and select new or press the new file icon on the toolbar as shown in Fig. (3.4).

- To print a file either go to the File menu and select print or press the icon for print on the toolbar as shown in Fig. (3.4).

- To solve, go to Solver menu and select solve or press the icon for solving on the toolbar as shown in Fig. (3.4).

- Some terms appear in "Solution Report-LINGO1" such as:
  Reduced Cost: The amount by which the objective function can improve to reach profitability.
  Slack/Surplus: Tells how close the solution is to satisfy a constraint as an equality.
  Dual Price: The amount that the objective would improve as the RHS or constant term in the constraint is increased by one unit. These are also called as "Shadow Prices".

3.1.5 Logical Operators and Set Looping Functions

The description and signs of some logical operators and set looping functions used in LINGO are shown in Tables (3.2) and (3.3).

Table 3.2: Logical Operators.

Operator	Sign
EQ	$=$
NE	$\neq$
GE	$\geq$
GT	$>$
LT	$<$
LE	$\leq$

Table 3.3: Set Looping Functions.

Looping Functions	Description
@FOR	Generate constraints over set members and to assign values to attributes.
@SUM	Computes the sum of an expression over all members of a set.
@MAX	Computes the maximum of an expression over all the members of a set.
@MIN	Computes the minimum of an expression over all the members of a set.
@PROD	Computes the product of an expression over all the members of a set.

3.1.6 Variable Domain Functions

The description of variable domain functions used in LINGO are given in Table (3.4):

Table 3.4: Variable Domain Functions.

Domain Functions	Description
@GIN	restricts a variable to being an integer value
@BIN	makes a variable binary (i.e., 0 or 1) (assignment problems)
@FREE	allows a variable to assume any real value, positive or negative (in case of forecasting examples)
@BND	limits a variable to fall within a finite range
@SOS	defines a set of binary variables and places restrictions on their collective values
@CARD	defines a set of binary variables and places an upper limit on their sum
@SEMIC	restricts variables to being either zero or greater-than a specified constant
@PRIORITY	used to assign branching priorities to variables
@POSD	restricts a square matrix to being a symmetric matrix and a positive semi-definite

Examples for the use of some domain functions are as follows:

```
! Here is the total profit objective function;
MAX = 100 * PRODUCT1 + 150 * PRODUCT2;
! Constraints on the production line capacity;
PRODUCT1<=103;
PRODUCT2<= 120;
! Our labor supply is limited;
PRODUCT1 + 2 * PRODUCT2 <= 160;
! Integer values only;
@GIN( PRODUCT1); @GIN( PRODUCT2);
```

Similarly, if the manufacturer wants to produce only one product and wants to know which product will maximize the profit, then @BIN domain function can be used, and if the manufacturer wants to limit the products, then @BND domain function is used.

```
! Here is the total profit objective function;
MAX = 100 * PRODUCT1 + 150 * PRODUCT2;
! Constraints on the production line capacity;
PRODUCT1<=103;
PRODUCT2<= 120;
! Our labor supply is limited;
PRODUCT1 + 2 * PRODUCT2 <= 160;
! Integer values only;
@BIN( PRODUCT1); @BIN( PRODUCT2);
```

3.1.7 Shortcut Keys

Some useful shortcut keys are given in the Table (3.5) below:

Table 3.5: Shortcut Keys.

Operation	Shortcut Key		
New Model Window	F2	Open Existing File	Ctrl+O
Save	Ctrl+s	Save As	F5
Close Active Window	F6	Print	F7
Print Setup	F8	Print Preview	Shift+F8
Log Output	F9	Import LINDO File	F12
Exit	F10	Undo	Ctrl+Z
Cut	Ctrl+X	Copy	Ctrl+C
Paste	Ctrl+V	Select All	Ctrl+A
Find	Ctrl+F	Find Next	Ctrl+N
Replace	Ctrl+H	Go To Line	Ctrl+T
Match Parenthesis	Ctrl+P	Solve	Ctrl+U
Solution	Ctrl+W	Range	Ctrl+R
Options	Ctrl+I	Generate	Ctrl+G/Ctrl+Q
Picture	Ctrl+K	Model Statistics	Ctrl+E
Look	Ctrl+L	Command Window	Ctrl+1
Status Window	Ctrl+2	Close All	Ctrl+3

OPTIMIZATION II

Chapter 4
Introduction to Optimization

4.1 Introduction

The core concept of operations research is optimization, which is all about making better and more informed decisions. All of us are confronted with making decisions daily. Some decisions impact and shape our lives, while others hardly have any impact. Nevertheless, making the right choices is a fundamental life skill, relevant to everyone. Optimization can be done with or without limitations. Numerous optimization functions exist in real-life application. They include linear, non-linear, stochastic, fuzzy, and many others. These could be single or multiple objectives in nature. The details of such types of optimization are discussed with applications in the next chapter. This chapter discusses the prerequisite theorems and concepts with examples.

4.1.1 Unconstrained vs Constrained

An optimization problem is of two types:

- Unconstrained Optimization
- Constrained Optimization

If no constraints/limitations are specified or imposed on the objective function, then the problem is referred to as an Unconstrained Optimization Problem. On the other hand, when the optimization problem(s) involves optimizing the objective function or goal under the set of constraints, then that problem is known as

Constrained Optimization Problem. Lets define the general mathematical form of constrained optimization problem:

$$\left.\begin{array}{l}\min(\max) f(\underline{x}) \\ \text{subject to } g_i(\underline{x}) \leq 0 \\ \phantom{\text{subject to }} h_i(\underline{x}) = 0 \\ \underline{x} \geq 0,\ \underline{x} \in R^n\end{array}\right\} \quad (4.1)$$

The optimum vector $\underline{x}$ that solves the problem Eqn. (4.1) is denoted by $\underline{x}^*$ with corresponding optimum function value $f(\underline{x}^*)$. The problem defined in Eqn. (4.1) is referred to as a constrained optimization problem.

In unconstrained optimization problems, the aim is to optimize $f(\underline{x})$ and obtain $X^* = (x_1, x_2, \ldots, x_n)^T$ without any restrictions. Whereas in constrained optimization, the $f(x)$ is optimized to obtain $X^* = (x_1, x_2, \ldots, x_n)^T$ subject to the inequality constraints such as

$(i)\ g_i(\underline{x}) \leq 0, \quad (ii)\ h_i(\underline{x}) \geq 0, \quad and \ l_i(\underline{x}) = 0$

4.1.2 Linear vs Non-Linear

The problem defined in Eqn. (4.1) is referred to as a Linear optimization problem if the objective function $f(\underline{x})$ and defined constraints $g_i(\underline{x})$ and $h_i(\underline{x})$ all are linear. The detailed linear optimization has been discussed in Chapter 5. Problem (4.1) is referred to as a non-linear if either objective function or constraints or both are non-linear. Chapter 6 presents the in-depth insights of non-linear optimization problems.

In the next section, the definitions of basic concepts behind the optimization are discussed.

4.2 Basic Definitions

Some basic concepts are defined here for a quick understanding of the topic.

4.2.1 Affine Sets

Definition 4.1 A set $C \subseteq \mathbb{R}^n$ is affine if the line through any two distinct points in C lies in C. That is, if for any $x_1, x_2 \in C$ and $\theta \in R$, we have $\theta x_1 + (1-\theta)x_2 \in C$.

Let C be an affine set, $x_1, x_2, \ldots, x_k \in C$, and $\theta_1 + \theta_2 + \ldots + \theta_k = 1$, then the point $\theta_1 x_1 + \theta_2 x_2 + \ldots + \theta_k x_k$ also belongs to C.

4.2.2 Convex Set

Definition 4.2 A set C is convex iff $x, y \in C$

$$\Rightarrow \alpha x + (1-\alpha)y \in C \quad \forall \; \alpha \in [0,1]$$

For a collection of convex sets, the other convex sets can be created by the operations of dilation, sum, and intersection.

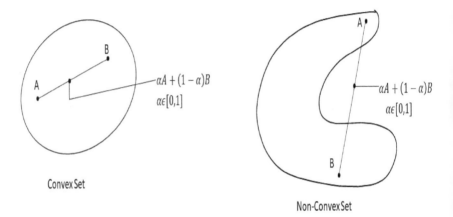

Figure 4.1: Convex and Non-Convex Sets.

Property

1. If A and B are convex sets then the dilation λA of a convex set A is convex. That is, $\lambda A = \{\lambda x : x \in A\}$ is convex.

2. The sum $A + B$ of convex sets is convex. That is, $\{a + b : a \in A, b \in B\}$ is convex.

3. The arbitrary intersection of convex sets is convex.

4.2.3 Convex Function

Definition 4.3 A function is convex if and only if

$$f(\alpha x + (1-\alpha)y) \leq \alpha f(x) + (1-\alpha)f(y) \quad \forall \; \alpha \in [0,1]$$

A function f is said to be strictly convex if, for any two distinct points $\underline{x}$ and $\underline{y}$ and for every scalar α, where $\alpha \in [0,1]$,

$$f(\alpha x + (1-\alpha)y) < \alpha f(x) + (1-\alpha)f(y)$$

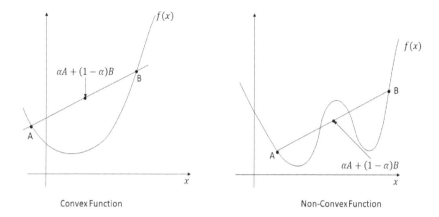

Figure 4.2: Convex and Non-Convex Functions.

Definition 4.4 A point $x \in X$ is an extreme point of the convex set X iff $\alpha x_1 + (1-\alpha)x_2$, $0 < \alpha < 1$; $x_1 \neq x_2$.

Remarks:

1. Any extreme point is on the boundary of the set.
2. Not all boundary points of a convex set are necessarily extreme points. Some boundary points may lie between two other boundary points.
3. Convex sets in R^n satisfy the following properties
   (i) If X is a convex set and $\beta \in R$, the set $\beta X = \{y : y = \beta x, x \in X\}$ is convex.
   (ii) If X and Y are convex sets, then the set $X + Y = \{Z : Z = (x+y), x \in X, y \in Y\}$ is convex.
   (iii) The intersection of any collection of convex sets is convex.
   (iv) Any local minimum of a convex function $f(x)$ is a global minimum.
   (v) A linear function is both convex and concave on R^n.
   (vi) If $f_i(x)$, $i = 1, 2, \ldots, n$ are convex functions defined on a convex set X and $\lambda_i \geq 0$; $i = 1, 2, \ldots, n$, then $\sum_{i=1}^{n} \lambda_i f_i(\underline{x})$ is also a convex function on X.

4.2.4 Concave Function

Definition 4.5 A function is said to be concave over a convex set X if, for any two points x_1 & $x_2 \in X$ $\forall$ $\lambda \in [0,1]$, with $\hat{x} = \lambda x_2 + (1-\lambda)x_1$,

$$f(\hat{x}) \geq \lambda f(x_2) + (1-\lambda)f(x_1) = g(\hat{x})$$

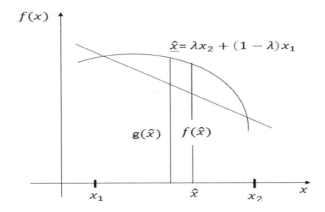

Figure 4.3: Concave Function.

In the Fig. (4.3) the line segment joining the two points lie entirely below or on the graph of $f(x)$.

Remarks:

1. The function $f(x)$ is strictly concave if a strict inequality holds.

2. A concave function bends downwards, and hence the local maximum will also be its global maximum.

3. It is also observed that the negative of a convex function is a concave function and vice versa. That is, the function $f(x)$ is called concave, if $-f(x)$ is convex.

4. A linear function can be both convex and concave as it satisfies conditions of both convexity and concavity functions.

5. The sum of convex functions and concave functions is also convex and concave, respectively.

Definition 4.6 Let $S \subset R^n$. The convex hull of S is the set, which is the intersection of all convex sets containing S.

Example 46 *Is the function $f(x) = 2x + 3$ convex or concave?*

$$\frac{df(x)}{dx} = +2$$

$$\frac{d^2 f(x)}{dx^2} = 0$$

Hence, the function is both convex and concave.

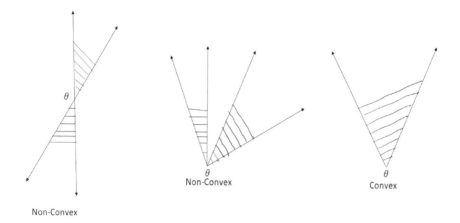

Figure 4.4: Convex and Non-Convex Sets.

Example 47 *Is the function* $f(x) = 2x^3 - x^2 + 4x + 8$ *convex or concave?*

$$\frac{df(x)}{dx} = 6x^2 - 2x + 4$$

$$\frac{d^2f(x)}{dx^2} = 12x - 2$$

Since $\frac{d^2f(x)}{dx^2}$ is not greater than or equal to (or less than or equal to) zero for x from $-\infty$ to $+\infty$, the function is neither convex nor concave.

Example 48 *Is the function* $f(x) = 2x^2 - 4x + 5$ *convex or concave?*

$$\frac{df(x)}{dx} = 4x - 4$$

$$\frac{d^2f(x)}{dx^2} = +4 > 0$$

Hence, the function is strictly convex.

4.3 Convex and Concave Properties

1. Function e^{ax} is convex on $\mathbb{R}$, for any $a \in \mathbb{R}$.
2. Function x^a is convex on $\mathbb{R}++$, when $a \geq 1$ or $a \leq 0$, and concave for $0 \leq a \leq 1$.
3. Function $|x|^p$, for $p \geq 1$, is convex on $\mathbb{R}$.
4. Function $\log x$ is concave on $\mathbb{R}++$.

5. Every norm on $\mathbb{R}^n$ is convex.
6. Maximum function $f(x) = \max\{x_1, x_2, \ldots, x_n\}$ is convex on $\mathbb{R}^n$.
7. Quadratic over linear function $f(x,y) = \frac{x^2}{y}$ is convex.
8. Function $f(x) = \log(e^{x_1} + e^{x_2} + \ldots + e^{x_n})$ is convex on $\mathbb{R}^n$.
9. If f is convex, then $\exp f(x)$ is convex.
10. If f is concave and positive, then $\log f(x)$ is concave.
11. If f is concave and positive, then $1/f(x)$ is convex.
12. If f is concave and non-negative, then $f(x)^p, p \geq 1$ is convex.

4.4 Convex Optimization Problems

A convex optimization problem in standard form is defined as:

$$\begin{aligned}
\min \quad & f_0(x) \\
\text{subject to} \quad & f_i(x) \leq 0, \ i = 1, 2, \ldots, m \\
& a_i' x = b_i, \ i = 1, 2, \ldots, p
\end{aligned} \quad (4.2)$$

The convex problem (4.2) has three basic requirements:

i the objective function must be convex,

ii the inequality constraint functions must be convex,

iii the equality constraint functions $h_i(x) = a_i' x - b_i$ must be affine.

Also, the feasible set of a convex optimization problem is convex, since m convex sets $\{x | f_i(x) \leq 0\}$ and p hyperplane $\{x | a_i' x = b_i\}$. Noted that, it is assumed that $a_i \neq 0$. If $a_i = 0$, and $b_i = 0$ for some i, then the ith equality constraints can be deleted. If $a_i = 0$, and $b_i \neq 0$, the ith equality constraints are inconsistent, and the problem becomes infeasible.

In convex optimization problems, we minimize a convex objective function over a set of convex constraints. If the objective function $f_0(x)$ is *quasi convex* in problem (4.2), then it is referred to as a *quasi convex* optimization problem.

4.5 Concave Maximization Problem

Let us consider a mathematical programming problem:

$$\begin{aligned}
\max \quad & f_0(x) \\
\text{subject to} \quad & f_i(x) \leq 0, \ i = 1, 2, \ldots, m \\
& a' x = b_i, \ i = 1, 2, \ldots, p
\end{aligned} \quad (4.3)$$

The objective function in problem (4.3) is concave, while the inequality constraints are convex. The problem can be easily converted into convex optimization using the property $\min -f_0(x)$ (convex objective function). Hence, we can use all results, properties, and algorithms describing convex optimization problems. Similarly, problem (4.3) is called *quasi concave* if f_0 is quasi concave.

Global Minima

A point x_0 is said to be a *global minima* of $f(x)$ over the close set $X \in E^n$, if for all $x \in X$
$$f(x) \geq f(x_0).$$

Global Maxima

A point x_0 is said to be a *global maxima* of $f(x)$ over the close set $X \in E^n$, if for all $x \in X$
$$f(x) \leq f(x_0).$$

Local Minima

A point x_0 is said to be a *local minima* of $f(x)$ if there exist an ε *neighbourhood* of x_0, $\varepsilon > 0$ such that
$$f(x) \geq f(x_0).\ \forall \{x| \|x - x_0\| < \varepsilon\}.$$
For strong local minima, $f(x) > f(x_0).\ \forall \{x| \|x - x_0\| < \varepsilon\}$.

Local Maxima

A point x_0 is said to be a *local maxima* of $f(x)$ if there exists a neighbourhood ε of x_0, $\varepsilon > 0$ such that
$$f(x) \leq f(x_0).\ \forall \{x| \|x - x_0\| < \varepsilon\}.$$
For strong local maxima, $f(x) < f(x_0).\ \forall \{x| \|x - x_0\| < \varepsilon\}$.

4.5.1 Quasi Concave & Quasi Convex Function

A function f on a convex set S of a linear space X is quasi convex if
$$f(\alpha x_1 + (1-\alpha)x_2) \leq \max\{f(x_1), f(x_2)\}\ \forall\ x_1, x_2 \in S \text{ and } 0 \leq \alpha \leq 1$$
Similarly, f is quasi concave if
$$f(\alpha x_1 + (1-\alpha)x_2) \geq \min\{f(x_1), f(x_2)\}\ \forall\ x_1, x_2 \in S \text{ and } 0 \leq \alpha \leq 1$$
It is strictly quasi concave if the inequality is strict, that is, for every $x_1 \neq x_2$
$$f(\alpha x_1 + (1-\alpha)x_2) > \min\{f(x_1), f(x_2)\}\ 0 \leq \alpha \leq 1$$

It is noted that, geometrically a function is quasi concave if the function along a line joining any two points in the domain lies above at least one of the end points.

Note:

1. f is quasi concave iff $-f$ is quasi convex.

2. Every concave function is quasi concave.

3. Any monotonic function on $\mathbb{R}$ is both quasi convex and quasi concave. For example-$f(x) = x^a$ at $\mathbb{R}$ is monotonic, if $a \geq 0$ function is strictly increasing and strictly decreasing for $a \leq 0$. Therefore, it is quasi concave (and quasi convex) $\forall\ a$.

4. A function f is quasi concave iff every upper contour set is convex.

5. A function f is quasi convex iff every lower contour set is convex.

4.5.1.1 Properties of Quasi Concave Functions

1. Let function $f(x)$ be concave and $g(x)$ be quasi concave, but their sum $(f+g)x$ is neither concave nor quasi concave.

2. If function $f(x)$ is quasi concave and $g(x)$ is increasing, then $g \circ f$ is quasi concave.

3. Let $f(x)$ and $g(x)$ be affine functions on a linear space X, and let $S \subseteq X$ be a convex set on which $g(x) \neq 0$. The function $h(x) = \frac{f(x)}{g(x)}$ is both quasi concave and quasi convex on S.

4. Let f and g be strictly positive definite functions on a convex set S with f concave and g convex. Then, $h(x) = \frac{f(x)}{g(x)}$ is quasi concave on S.

5. Let f and g be strictly positive definite concave functions on a convex set S. Then, $h(x) = f(x)\dot{g}(x)$ is quasi concave on S.

6. If f is non-negative definite, then $\log f$ concave $\Rightarrow f$ is quasi concave.

7. Let $f_1, f_2, \ldots, f_n$ be non-negative definite concave functions on a convex set S. Then $f(\underline{x}) = (f(x_1))^{\alpha_1} (f(x_2))^{\alpha_2} \ldots (f(x_n))^{\alpha_n}$ is quasi concave on S for any $\alpha_1, \alpha_2, \ldots, \alpha_n \in R$

Theorem 4.1
Convex Maximum Theorem: *Assume that $f : X \times \Theta \to R$ is convex in θ. Then the value function $V(\theta) = \max_{x \in X} f(x, \theta)$ is convex in θ.*

Theorem 4.2
Concave Maximum Theorem: *Consider the general constrained maximization problem:*

$$\max_{x \in G(\theta)}$$

where X and Θ are linear spaces. Let $\Theta^* \subset \Theta$ denote the set of parameter values for which a solution exists if

(i) *the objective function $f : X \times \Theta \to R$ is quasi concave in X and*

(ii) *the constraint corresponding $G : \Theta \to X$ is convex valued then the solution correspondence $\varphi : \Theta^* \to X$ defined by $\varphi(\theta) = \arg\ \max_{x \in G(\theta)} f(x, \theta)$ is convex valued. Furthermore, if*

(iii) *the objective function $f : X \times \Theta \to R$ is strictly concave in $X \times \Theta$ and*

(iv) *the constraint correspondence $G : \Theta \to X$ is convex the value function $V(\theta) = \sup_{x \in G(\theta)} f(x, \theta)$ is strictly concave in θ.*

Corollary: Let $\Theta^* \subset \Theta$ denote the set of parameter values for which a solution exists in the general constrained maximization problem $\max_{x \in G(\theta)} f(x, \theta)$, where X and Θ are linear spaces. If

(i) the objective function $f : X \times \Theta \to R$ is strictly quasi concave in X and

(ii) the constraint correspondence $G : \Theta \to X$ is convex valued then the solution correspondence $\varphi : \Theta^* \to X$ defined by $\varphi(\theta) = \arg\ \max_{x \in G(\theta)} f(x, \theta)$ is single valued; that is, φ is a function from Θ^* to X.

Example 49 *Suppose the constrained optimization problem with the constraint set $G(\theta)$*

$$\left. \begin{aligned} \max_{x \in G(\theta)}\ & f(x, \theta) \\ \text{subject to}\ & g_1(x, \theta) \leq b_1 \\ & g_2(x, \theta) \leq b_2 \\ & \quad \vdots \\ & g_m(x, \theta) \leq b_m \end{aligned} \right\} \quad (4.4)$$

where each $g(x, \theta)$ is convex jointly in x and θ. Then, the correspondence $G(\theta) = \{x \in X : g_j(x, \theta) \leq b_j,\ j = 1, 2, \ldots, m\}$ is convex. Provided that the function $f(x, \theta)$ is (strictly) concave in x and θ, the value function $V(\theta) = \sup_{x \in G(\theta)} f(x, \theta)$ is (strictly) convex in θ (see concave maximum theorem).

4.6 Convexity of Smooth Function

Theorem 4.3
If $f(x)$ is a differentiable function over the convex set $X \subseteq R^n$ then $f(x)$ is convex over X iff
$$f(x_2) - f(x_1) \geq \nabla' f(x_1)(x_2 - x_1). \forall x_1, x_2 \in X.$$
It is noted, if $f(x)$ is to be strictly convex, then
$$f(x_2) - f(x_1) > \nabla' f(x_1)(x_2 - x_1). \forall x_1, x_2 \in X.$$

Theorem 4.4
If $f(x)$ is a twice continuously differentiable function (i.e., $f(x) \subset C^2$) over an open convex set $X \subseteq R^n$, then $f(x)$ is convex iff $H(x)$ (Hessian matrix) is a positive semi definite $\forall x \in X$.

Note that, if in the theorem (4.4) above, Hessian matrix is positive definite $\forall x \in X$, then $f(x)$ is strictly convex over X.

Theorem 4.5
If $f(x)$ is a twice continuously differentiable function over an open convex set $X \subseteq R^n$, and if $f(x)$ is convex over X, then any interior local minima of $f(x)$ is a global minima.

Note the following:

1. Function $f(x)$ is concave, iff $f(x_2) - f(x_1) \leq \nabla' f(x_1)(x_2 - x_1). \forall x_1, x_2 \in X$, and strictly concave, iff $f(x_2) - f(x_1) < \nabla' f(x_1)(x_2 - x_1). \forall x_1, x_2 \in X$.

2. Geometrically, a differentiable function is convex iff it lies above its tangent hyperplane at every point in the domain.

3. A differentiable function f on a convex, open set $X \in R^n$ is convex iff
$$[\nabla f(x_2) - \nabla f(x_1)]' (x_2 - x_1) \geq 0. \forall x_1, x_2 \in X.$$
Function f is strictly convex iff
$$[\nabla f(x_2) - \nabla f(x_1)]' (x_2 - x_1) > 0. \forall x_1, x_2 \in X.$$

4. A smooth function f on an open interval $X \subseteq \mathbb{R}$ is strictly convex iff f' is (strictly) increasing.

5. If f is twice differentiable function on an open interval $X \subseteq \mathbb{R}$. Then, f is convex iff $f''(x) \geq 0$, $\forall x \in X$. And f is concave iff $f''(x) \leq 0$, $\forall x \in X$. f is strictly convex iff $f''(x) > 0$, $\forall x \in X$. And f is strictly concave iff $f''(x) < 0$, $\forall x \in X$.

4.6.1 Quasi Concavity of Smooth Function

A differentiable function is quasi concave iff its upper contour sets lie above the tangent to the contour.

A differentiable function f on an open set $X \subseteq \mathbb{R}^n$ is quasi concave iff

$$f(x_2) \geq f(x_1) \implies \nabla' f(x_1)(x_2 - x_1) \geq 0, \forall x_1, x_2 \in X \qquad (4.5)$$

The inequality can hold strongly where f is regular. At every regular point $\nabla f(x_1) \neq 0$,

$$f(x_2) > f(x_1) \implies \nabla' f(x_1)(x_2 - x_1) > 0, \forall x_1, x_2 \in X \qquad (4.6)$$

A function is a quasi concave if it satisfies Eqn. (4.6) at a regular point of f.

4.6.2 Pseudo Concave Function

A function is said to be Pseudo concave if it satisfies Eqn.(4.6) at all points of its domain.

A differentiable function f on an open convex set $X \subseteq \mathbb{R}^n$ is a Pseudo concave if

$$f(x_2) > f(x_1) \implies \nabla' f(x_1)(x_2 - x_1) < 0, \ \forall x_1, x_2 \in X \qquad (4.7)$$

Pseudo concave functions have two advantages over quasi concave functions.

i Every local optima is a global optima.

ii The second derivative test for pseudo concave functions is easier than quasi concave.

4.6.3 Pseudo Convex Function

A differentiable function $f: X \longrightarrow R$ is pseudo convex if

$$f(x_2) < f(x_1) \implies \nabla' f(x_1)(x_2 - x_1) > 0, \ \forall x_1, x_2 \in X \qquad (4.8)$$

A function is pseudo convex if $-f$ is pseudo concave.

Note the following:

1. In the constrained optimization problem

$$\max_{x \in X} f(x) \qquad (4.9)$$

If f is concave and X convex, then every local optima is a global optima.

2. In the constrained optimization problem (4.9), if f is strictly quasi concave and X convex, then every optima is a global optima.

4.6.4 Cone

Definition 4.7 A cone C is a set such that if $x \in C$, then $\theta x \in C$, $\forall \theta \in R_+$. A cone which is also convex is known as a convex cone. Let x_1, x_2 be any points $\in C$ and $\theta_1, \theta_2 \geq 0$, we have
$$\theta_1 x_1 + \theta_2 x_2 \in C.$$
A point of the form $\theta_1 x_1 + \theta_2 x_2 + \ldots + \theta_k x_k$ with $\theta_1, \ldots, \theta_k \geq 0$ is called a conic combination or a non negative linear combination of $x_1, \ldots, x_k$. If all x_i are in the convex cone C, then every conic combination of x_i is in C. In other words, we can say, a set C is a convex cone iff it contains all conic combinations of its elements.

Definition 4.8 A conic hull of a set C is the set of all conic combinations of points in C. That is,
$$\{\theta_1 x_1 + \theta_2 x_2 + \ldots + \theta_k x_k | x_i \in C, \ \theta_i \geq 0, \ \forall \, i = 1, 2, \ldots, k\}$$

4.6.5 Hyperplanes and Half Spaces

Definition 4.9 A hyperplane is a set of the form $\{x | a^T x = b\}$, where $x \in R^n$, $a \neq 0$, and $b \in R$.

The hyperplane $\{x | a^T x = b\}$ is the set of points with normal vector a; the constant $b \in R$ determines the offset of the hyperplane from the origin. That is,
$$\{x | a^T (x - x_0) = 0\}$$
where x_0 is any point in the hyperplane, that is, any point that satisfies $a^T x_0 = b$. A hyperplane divides R^n into two half spaces. The half space determined by $a^T x \geq b$ is the half space extending in the direction a.

The half space determined by $a^T x \leq b$ extends in the direction $-a$, where $a \neq 0$, half spaces are convex but not affine.

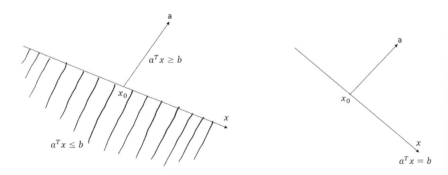

Figure 4.5: Hyperplanes.

The boundary of the (closed) half space $\{x|a^T x \leq b\}$ is the hyperplane $\{x|a^T x = b\}$. The set $\{x|a^T x < b\}$, which is the interior of the half space $\{x|a^T x \leq b\}$, is called an open half space.

4.7 Univariate Optimization Problems

Definition 4.10 If a function (min or max) has only one variable which is to be optimized then the problem is referred as a single variable optimization problem. If certain conditions are imposed on the function then the problem becomes a one variable constrained optimization problem.

Suppose a single real variable function $f(x)$ is continuous and a twice continuous differentiable function, that is, $\min f(\underline{x})$, $\underline{x} \in R$. For a strong local minimum, we search for x^* for which $f(x^*) < f(x)$ $\forall x$.

As we know that x^* occurs where the slope is zero, that is, $f'(\underline{x}^*) = 0$, which is a first order necessary condition. It is also required that the second order condition $f''(\underline{x}^*) > 0$ hold at $\underline{x}^*$ for a strong local minimum.

Example 50 *Consider a quadratic function $f(x) = ax^2 + bx + c$, for which the minimum occurs where $f(x) = 0$, that is $\underline{x}^* = -\frac{b}{2a}$, provided $f''\underline{x}^* = a > 0$.*

It is noted that if the function $f(x)$ is to be maximized in the whole real domain R, that is, $\max f(\underline{x})$, $\underline{x} \in R$. For a strong local maximum, we search for x^* for which $f(x^*) > f(x)$ $\forall x$ and x^* occurs where $f'(x) = 0$ and $f''(x) < 0$. However, if $f'(x) = 0$ but $f''(x)$ is indefinite then x^* corresponds to a saddle point. In general, a function can have several stationary points. Therefore, to identify the global maximum or minimum, we need to check every stationary point, unless the objective function is convex.

Some established methods for solving single variable optimization problems include: Unrestricted search, Method of golden section, Fibonacci search, Quadratic interpolation, and Newton-Raphson method, amongst others.

4.7.1 Optimality Condition

A function $f(x)$ attains it's maximum or minimum value at $x = x_0$ in the interval $a \leq x \leq b$ if the following conditions hold:

(i) $f'(x_0) = 0$ (necessary condition): It is the necessary condition for any function $f(x)$ to have a local optimum value at any extreme point $x = x_0$.

(ii) $f'(x_0) = f''(x_0) = \ldots = f^{n-1}(x_0) = 0$ and $f^n(x_0) \neq 0$ (necessary condition)
Based on this condition, we have the following conclusions:

   a) If $f^n(x_0) < 0$, for an even n, then the function has a local maxima at x_0.

b) If $f^n(x_0) > 0$, for an even n, the function has a local minima at x_0.

c) If $f^n(x_0) \neq 0$, for an odd n, the function has a point of inflection at x_0.

Example 51 *Consider the constrained optimization problem.*

$$\left. \begin{array}{l} \min f(x_1,x_2) = (x_1-3)^2 + (x_2-4)^2 \\ \text{subject to } 2x_1 + x_2 = 7 \end{array} \right\} \quad (4.10)$$

The unconstrained minimum occurs at (3,4) and the function value $Z = 0$. The constrained minimum is found by the substitution method, that is, $x_2 = 7 - 2x_1$, and the problem is reduced to

$$\min f(x_1) = (x_1-3)^2 + (7-2x_1-4)^2$$
$$= 5x_1^2 - 18x_1 + 18$$

Now,

$$f'(x_1) = 10x_1 - 18 = 0 \Rightarrow x_1 = 18/10$$

and

$$f''(x_1) = 10 \geq 0,$$

the global minimum occurs at $x_1^* = 18/10$ and $x_2^* = 34/10$. The optimal value of the objective function is $f^*(x_1,x_2) = 18/10$.

Example 52

$$\left. \begin{array}{l} \min f(x_1,x_2) = x_1^2 - 4x_1x_2 + 5x_2^2 + 2x_1 - 6x_2 \\ \text{subject to } x_1^2 + x_2^2 \leq 9 \\ \qquad\qquad x_1 + x_2 \leq 3 \\ \qquad\qquad x_1, x_2 \geq 0 \end{array} \right\} \quad (4.11)$$

The first constraint requires that a convex function be less than or equal to a constant, and the set of points satisfying it constitutes a convex set. The second constraint being linear, also defines convex sets, and the feasible region must itself be convex.

The objective function can be written as

$$f(x_1,x_2) = (x_1-2x_2)^2 + x_2^2 + 2x_1 - 6x_2$$

and since the quadratic portion is positive definite, the problem is evidently one of minimizing a convex function over a convex set. Check if there are any unconstrained local minima of $f(x_1,x_2)$ in the feasible region.

$$\frac{\partial f}{\partial x_1} = 2x_1 - 4x_2 + 2 \quad (4.12)$$

and

$$\frac{\partial f}{\partial x_2} = -4x_1 + 10x_2 - 6 \quad (4.13)$$

Both constraints in Eqns. (4.12) & (4.13) are zero at the point (1,1). Also, this point satisfies all the other constraints, and a check of the second derivative shows that it is an unconstrained local minimum, therefore, a constrained local minimum. Then, it must be the (global) optimal solution to the problem.

4.8 Multivariate Optimization Problems

In many situations a function may have several variables, its optimization can be expressed in a similar fashion to a univariate case.

$$\min(or \ \max) f(\underline{x}), \ \underline{x} \in R^n$$

For this case, the vector $\underline{x}$ is in an n-dimensional space where each component x_i is a real number.

4.8.1 Optimality Condition

Consider that the multivariate function $f(x,y)$ attains it optima (max/min) value if the following conditions hold:

(i) For relative maxima

    a) $f_x, f_y = 0$
    b) $f_{xx}, f_{yy} > 0$
    c) $f_{xx} \cdot f_{yy} - (f_{xy})^2 > 0$, since $f_{xy} = f_{yx}$ by young's theorem $f_{xy} \cdot f_{yx} = (f_{xy})^2$

(ii) For relative minima

    a) $f_x, f_y = 0$
    b) $f_{xx}, f_{yy} < 0$
    c) $f_{xx} \cdot f_{yy} - (f_{xy})^2 > 0$

It is noted that, if $f_{xx} \cdot f_{yy} < (f_{xy})^2$, when f_{xx} and f_{yy} have the same signs, the function is at an inflection point; when f_{xx} and f_{yy} have different signs the function is at a saddle point. If $f_{xx} \cdot f_{yy} = (f_{xy})^2$, no conclusion can be drawn regarding the function optima.

If the function is strictly convex (concave) in x and y, the function will have only one minima (maxima), called absolute or global minima (maxima). If the function is merely convex (concave) at x and y defined in the interval, the critical point is referred to as a relative or local minima (maxima).

Example 53 Consider the function $f(x,y) = 2y^3 - x^3 + 147x - 54y + 12$
First derivatives:
$$f_x = -3x^2 + 147 = 0 \quad (4.14)$$
$$f_y = 6y^2 - 54 = 0 \quad (4.15)$$

Solving Eqns. (4.14) & (4.15), we have $x = \pm 7$, $y = \pm 3$; there are four distinct sets of critical points: (7,3), (7,-3), (-7,3) and (-7,-3).

Second derivatives:
$$f_{xx} = -6x \quad f_{yy} = 12y$$

The values of second derivatives at the critical points are:

S.No.	Critical Points	f_{xx}	f_{yy}
(i)	(7,3)	−42 < 0	36 > 0
(ii)	(7,-3)	−42 < 0	−36 < 0
(iii)	(-7,3)	42 > 0	36 > 0
(iv)	(-7,-3)	42 > 0	−36 < 0

Since the function has different signs for the first and fourth critical points, therefore, the function cannot be at a local maxima or minima at (7,3) and (-7,-3). The functions f_{xx} and f_{yy} are of different signs, $f_{xx} \cdot f_{yy} \not> (f_{xy})^2$, and hence the function has a saddle point.

In case (ii) the function has a positive value, the function may be at a local maxima at (7,-3). While in case (iii) function is negative, the function may be at a local minima at (-7,3), but the third condition must be verified first to ensure against the possibility of an inflection point.

Example 54 Consider function $f(x,y) = 3x^2 - xy + 2y^2 - 4x - 7y + 20$
First derivatives:
$$f_x = 6x - y - 4 = 0 \quad (4.16)$$
$$f_y = -x + 4y - 7 = 0 \quad (4.17)$$

Solving Eqns. (4.16) & (4.17), we have critical point $x = 1, y = 2$.

Second derivatives are $f_{xx} = 6, f_{yy} = 4$. The value of second derivatives at the critical point are
$$f_{xx}(1,2) = 6 \geq 0$$
$$f_{yy}(1,2) = 4 \geq 0$$

Since second order derivatives are positive, the function is possibly at a global minimum.

Again we calculate the partial derivative for the function.
$$f_{xy} = -1$$

$$f_{yx} = -1$$

The value of the functions at critical point are

$$f_{xy}(1,2) = -1 = f_{yx}(1,2)$$

Now, we check the condition

$$f_{xx}(1,2)f_{yy}(1,2) \geq [f_{xy}(1,2)]^2$$

$$6 \times 4 \geq (-1)^2$$

In view of conditions $f_{xx}f_{yy} \geq (f_{xy})^2$ and $f_{xx}, f_{yy} \geq 0$ the function is at a global minima at (1,2).

Example 55 Consider a function $f(x,y) = 60x + 34y - 4xy - 6x^2 - 3y^2 + 12$
First derivative

$$f_x = 60 - 4y - 12x = 0 \qquad (4.18)$$

$$f_y = 34 - 6y - 4x = 0 \qquad (4.19)$$

Solving Eqns. (4.18) & (4.19), we have critical point $x = 4, y = 3$.

Second derivatives,

$$f_{xx} = -12, \quad f_{yy} = -6$$

The value of second derivatives at the critical point are,

$$f_{xx}(4,3) = -12 < 0$$

$$f_{yy}(4,3) = -6 < 0$$

Since the second order derivatives are negative, the function possibly is at global maximum.

Again calculating the partial derivatives for the function

$$f_{xy} = -4 = f_{yx}$$

Now, we check the condition

$$f_{xy}(4,3) = -4 = f_{yx}(4,3)$$

$$f_{xx}(4,3)f_{yy}(4,3) \geq [f_{xy}(4,3)]^2$$

$$-12 \times -6 \geq (-4)^2$$

$$72 \geq 16$$

In view of the condition $f_{xx}f_{yy} \geq (f_{xy})^2$ and $f_{xx}, f_{yy} \leq 0$, the function is at a global maxima at (4,3).

Example 56 Consider a function $f(x,y) = 72x + 48y - 3x^2 - 2y^2 - 6xy + 10$
First derivative

$$f_x = 72 - 6x - 6y = 0 \qquad (4.20)$$

$$f_y = 48 - 4y - 6x = 0 \qquad (4.21)$$

Solving Eqns. (4.20) & (4.21), we have critical point $x = 0, y = 12$.

The second derivatives are,

$$f_{xx} = -6, \quad f_{yy} = -4$$

The value of the second derivatives at the critical points are,

$$f_{xx}(0,12) = -6 < 0$$

$$f_{yy}(0,12) = -4 < 0$$

Since the second order derivatives are negative at critical point (0,12), the function may be at a global maximum at (0,12).

Again calculate the partial derivatives for the function

$$f_{xy} = -6 = f_{yx}$$

and $f_{xy}(0,12) = -6 = f_{yx}(0,12)$

Now, we check the condition

$$f_{xx}(0,12) f_{yy}(0,12) \geq [f_{xy}(0,12)]^2$$

$$-6 \times -4 \not\geq (-6)^2$$

$$24 \not\geq 36$$

It can be concluded based on the condition that f_{xx} and f_{yy} have the same sign and $f_{xx} f_{yy} \leq (f_{xy})^2$, the function is at an inflection point at (0,12).

Example 57 Consider a function $f(x,y) = 5x^2 - 3y^2 - 30x + 7y + 4xy + 15$
First derivatives:

$$f_x = 10x + 4y - 30 = 0 \qquad (4.22)$$

$$f_y = 4x - 6y + 7 = 0 \qquad (4.23)$$

Solving Eqns. (4.22) & (4.23), we have critical point $x = 2, y = 5/2$.

The second derivatives are,

$$f_{xx} = 10, \quad f_{yx} = -6$$

and
$$f_{xx}(2,5/2) = 10 > 0$$
$$f_{yx}(2,5/2) = -6 < 0$$

The cross partial derivatives are,
$$f_{xy} = 4 = f_{yx}$$

Now, we check the condition
$$f_{xx}(2,5/2)f_{yy}(2,5/2) \geq [f_{xy}(2,5/2)]^2$$
$$10 \times -6 \not\geq (-4)^2$$
$$-60 < 16$$

It can be concluded based on the condition that f_{xx} and f_{yy} have different signs and $f_{xx}f_{yy} \leq (f_{xy})^2$, the function is at a saddle point.

4.9 Gradient Vector of $f(x)$

A function $f(x)$ is assumed to be a smooth function if it is twice continuously differentiable, that is $f(x) \in C^2$. At any point x, a vector of first-order partial derivatives is referred to as the gradient vector.

$$\nabla f(x) = \left[\frac{\partial f(x)}{\partial x_1} \cdots \frac{\partial f(x)}{\partial x_n} \right]^T \quad (4.24)$$

Note: The directional derivative at x in the direction θ for a small change $d\lambda$ is $\frac{df(x)}{d\lambda}|_\theta = \nabla' f(x) \cdot \theta$.

4.9.1 Hessian Matrix of $f(x)$

If $f(x) \in C^2$, then at the point x there exists a matrix of second order partial derivatives that is referred as Hessian matrix $H(x)$ which is a $n \times n$ symmetrical matrix.

$$H(x) = \frac{\partial^2 f(x)}{\partial x_i \partial x_j} = \begin{bmatrix} \frac{\partial^2 f(x)}{\partial x_1^2} & \cdots & \frac{\partial^2 f(x)}{\partial x_1 \partial x_n} \\ \frac{\partial^2 f(x)}{\partial x_2 \partial x_1} & \cdots & \frac{\partial^2 f(x)}{\partial x_2 \partial x_n} \\ & \vdots & \\ \frac{\partial^2 f(x)}{\partial x_n \partial x_2} & \cdots & \frac{\partial^2 f(x)}{\partial^2 x_n} \end{bmatrix} \quad (4.25)$$

The Hessian matrix is used to test multi-variable function optima points. As, we recall the first order conditions are, $f_x = f_y = 0$ (necessary condition). A sufficient condition for multi-variable function $f(x,y)$ to be at an optima are:

(i) $f_{xx}, f_{yy} > 0$ for minima
   $f_{xx}, f_{yy} < 0$ for maxima

(ii) $f_{xx} f_{yy} > (f_{xy})^2$

For this second order conditions we used the Hessian matrix. A Hessian matrix is a determinant composed of all the second order partial derivatives. That is,

$$H = \begin{vmatrix} f_{xx} & f_{xy} \\ f_{yx} & f_{yy} \end{vmatrix}, \text{ where } f_{xx} = f_{xy}$$

It is noted that different types of tests are used to identify local maxima or minima by examining the minors of the Hessian matrix.

a) <u>Hessian matrix positive definite</u>: If the first principal minor, $|H_1| = f_{xx} > 0$ and the second minor $|H_2| = f_{xx} \cdot f_{yy} - (f_{xy})^2 > 0$, then the function has a local minima.

b) <u>Hessian matrix negative definite</u>: If the first principal minor, $|H_1| = f_{xx} < 0$ and the second minor $|H_2| > 0$, then the function has a local maxima.

c) <u>Hessian matrix semi-definite (or indefinite)</u>: If the above conditions (a) & (b) do not meet, then the extreme point may be either a maxima or a minima or neither.

The condition for a multi-variable function to be a maximum or minimum is summarized and given below:

Necessary Condition	Sufficient Condition	Conclusion
$\nabla f(x_0) = 0$	$H(x_0)$ is +ve definite	local min at $x = x_0$
$\nabla f(x_0) = 0$	$H(x_0)$ is -ve definite	local max at $x = x_0$
$\nabla f(x_0) = 0$	$H(x_0)$ is indefinite	point of inflection at $x = x_0$

Example 58 Let $f(x,y) = x^2 + y^2$
$f_x = 2x = 0 \Rightarrow x = 0$
$f_y = 2y = 0 \Rightarrow y = 0$

Since the function has only one critical point, it may be either a max or min.

Example 59 $f(x,y) = xy - x^2 - y^2 - 2x - 2y + 4$
$f_x = y - 2x - 2 = 0$
$f_y = x - 2y - 2 = 0$

After solving the equations, we have $x = y = -2$ is the critical point.
$f_{xx} = -2$, $f_{yy} = -2$ and $f_{xy} = f_{yx} = 1$.
To test the second-order conditions, the Hessian matrix is used as:

$$H = \begin{vmatrix} -2 & 1 \\ 1 & -2 \end{vmatrix} = 4 - 1 = 3 > 0$$

Since $f_{xx} < 0$, $f_{xx}f_{yy} - (f_{xy})^2 > 0$, the Hessian is negative definite, and the function is maximum at the critical values.

Chapter 5

Linear Optimization Problems

5.1 Introduction

A special case of mathematical optimization is linear optimization or linear programming or in short LP. LP is used to optimize the linear function known as "Objective Function" (i.e., performance, cost, profit, time, return on investment, among others) subject to a set of linear equalities or inequalities known as "Constraints," on the use of limited resources (i.e., space, energy, time, material, labour, among others). The intersection of these linear equalities or inequalities form a feasible region, which is a convex polytope. Here, "linear" refers to the linear relationship between the variables; that is, if demand increases two times, then the profit also increases proportionally, while the meaning of "programming" is to convert the verbal problem and given data into a mathematical model.

George B Dantzing first developed the LP model during World War II to derive the solutions of US military and air force tactical and strategic problems. But, now the LP problems are widely used in the fields of human resources, finance, management, health care, education, military, agriculture, scheduling problems, transportation, and many more. Various software packages are available to solve complex LP models such as R, TORA, SAS, MATLAB, LINDO, and more. But the formulation of the LP model cannot be done by any of these softwares. For modelling, one has to practice and understand the problem clearly. The general structure and basic definitions of the LP model are discussed in the next section.

5.2 General Form of LP Model

Generally, the LP model consists of three components, which are described in the following subsections.

5.2.1 Components of LP Model

- **Decision Variables:** Decision variables are the quantities that are to be determined. Usually they are denoted by $x_1, x_2, \ldots, x_n$. All the decision variables are continuous, controllable, and non-negative.

- **Objective Function:** The function representing how the decision variables are to be optimized (minimized or maximized) is known as Objective Function. In other words, the objective function measures a system's performance; therefore, it is also known as a measure of performance.

- **Constraints:** Resources are always available in certain limitations (constraints). Therefore, constraints limit the objective function to a certain degree of achievement.

5.2.2 Assumptions

There are four major assumptions of LP models.

1. **Certainty:** In a LP model, it is assumed that all input information or parameters related to the decision variables such as availability of resources, the requirement of resources, profit, or cost must be precisely known and may be constant.

2. **Additivity:** The objective function value and the total amount of each resource consumed/supplied must be equal to the sum of the respective individual contributions of the decision variables. For example-Let, x and y be two finished products; the total profit earned by the company to sell them must be equal to the sum of the profits earned when these products are sold separately. Similarly, the number of resources used for making product x and y must be equal to the sum of resources consumed for making these products individually.

3. **Proportionality:** The amount of each resource consumed/supplied and its contribution to the profit or cost in the objective function must be directly proportional to the value of each decision variable. For instance, if the cost of production for one product is $500, then the cost of making 100 units of the same product is $500 \times 100 = \$50,000$.

4. **Divisibility:** Under this assumption, all decision variables are allowed to take fractional values.

5.2.3 General Mathematical Model

Consider the Linear Programming Problem (LPP) in the standard form as:

$$\left.\begin{array}{c} \max(\min)\ Z = \sum_{j=1}^{n} c_j x_j \\ \text{subject to } \sum_{j=1}^{n} a_{ij} x_j (\leq, =, \geq) b_i;\ i = 1, 2, \ldots, m \\ \text{and } x_j \geq 0 \end{array}\right\} \quad (5.1)$$

where
c_j represents the per unit cost (or profit or time) of the jth decision variable
x_j represents the jth decision variable
a_{ij} refers to the technological coefficients
b_i denotes the total availability of the ith resource
and among the signs of $\leq, =, \geq$ only one sign can hold for one constraint.

The LP model in (5.1) can be written in matrix form as:

$$\left.\begin{array}{lr} \max(\min)\ Z = \underline{c}\underline{x} & (i) \\ \text{subject to } A\underline{x}\ (\leq, =, \geq)\ \underline{b} & (ii) \\ \text{and } \underline{x} \geq 0 & (iii) \end{array}\right\} \quad (5.2)$$

Some basic definitions of problem (5.2) are as follows:

Definition 5.1 If a given vector $\underline{x} = (x_1, x_2, \ldots, x_n)$ satisfies the constraints (ii)-(iii) then vector $\underline{x}$ is a feasible solution of the LPP (5.2).

Definition 5.2 If a given vector $\underline{x} = (x_1, x_2, \ldots, x_n)$ is a feasible solution of the LPP (5.2) and provides the maximal (minimal) value for the objective function Z over the feasible set S, then the vector $\underline{x}$ is an optimal solution of the LPP (5.2).

Definition 5.3 A max(min) LPP is said to be solvable if its feasible set S is not empty, that is $S \neq \phi$, and objective function Z has a finite upper (lower) bound on S.

Definition 5.4 If the feasible set is empty, that is $S = \phi$; then the LPP is infeasible.

Definition 5.5 If objective function Z of an LPP has no upper (lower) finite bound, then the problem is unbounded.

Definition 5.6 The LPP is solvable, if
- feasible set S is not empty, that is, there exists at least one such vector $\underline{x}$ that satisfies constraints (ii)-(iii).
- objective function Z has a finite upper bound over set S.

In other cases, an LPP is said to be unsolvable.

Definition 5.7 The given vector $\underline{x} = (x_1, x_2, \ldots, x_n)^T$ is a basic solution to the system $A\underline{x} = b$, if vector $\underline{x}$ satisfies system constraints (ii) and (iii).

Definition 5.8 A point $\underline{x}$ in a convex set S is called an extreme point of S if x cannot be expressed as a convex combination of any other two distinct points of S.

Definition 5.9 A basic solution $\underline{x}$ is degenerate if at least one of its basic variables is equal to zero; otherwise, it is said to be non-degenerate.

Definition 5.10 A basic solution $\underline{x} = (x_1, x_2, \ldots, x_n)^T$ of system $A\underline{x} = b$ is said to be a basic feasible solution (BFS) of the LPP (5.2) if all the elements x_j, $j = 1, 2, \ldots, n$ of vectors x satisfy the non-negativity constraint (iii).

Theorem 5.1
The set of all feasible solutions to an LPP is a convex set.

Proof 5.1 Consider the LPP

$$\left.\begin{array}{c} \min\ Z = \underline{c}'\underline{x} \\ \text{subject to } A\underline{x} = \underline{b} \\ \text{and } \underline{x} \geq 0 \end{array}\right\} \quad (5.3)$$

The set of all feasible solutions denoted by

$$F = \{\underline{x} | A\underline{x} = \underline{b}, \underline{x} \geq 0\}$$

Let $\underline{x}_1$ and $\underline{x}_2 \in F$. Define the convex combination of $\underline{x}$ as

$$\underline{x} = \alpha \underline{x}_1 + (1-\alpha)\underline{x}_2,\ 0 \leq \alpha \leq 1$$

Then,

$$A\underline{x} = A(\alpha \underline{x}_1 + (1-\alpha)\underline{x}_2)$$
$$= \alpha A \underline{x}_1 + (1-\alpha)A\underline{x}_2$$
$$= \alpha \underline{b} + (1-\alpha)\underline{b}\ \because A\underline{x}_1 = \underline{b},\ A\underline{x}_2 = \underline{b}$$
$$= \underline{b}$$

Since $\underline{x}_1, \underline{x}_2, \alpha$, & $(1-\alpha)$ are all non-negative, therefore $\underline{x} \in F$ and hence F is convex.

5.2.4 Properties of Solution to an LPP

1. If there is a feasible solution to the system of constraints $A\underline{x} = \underline{b}$ and $\underline{x} \geq 0$, then there also exists a basic feasible solution to this system of constraints.

2. The objective function of an LPP assumes its minimum at an extreme point of the convex set $F = \{\underline{x} | A\underline{x} = \underline{b}, \underline{x} \geq 0\}$ generated by the set of all feasible solutions to the LPP. If it assumes its minimum at more than one extreme point, then it takes on the same value for every convex combination of those particular points.

3. The objective function of an LPP attains its minimum at one of the extreme points of F.

4. If $z_j - c_j \geq 0$ (optimality criteria), for any fixed j, then the given solution $\underline{x}_0$ (say) can always be improved for the LPP (5.3).

5. If for any basic feasible solution $\underline{x}_0$, the optimality condition $z_j - c_j \leq 0$ for all j, then $\underline{x}_0$ will be the required optimal solution to the LPP (5.3).

5.3 Formulation of an LPP

The formulation of a LPP involves the following steps:

Step 1 Identify the decision variables of the problem.

Step 2 Frame the objective function of the problem, which is a linear function of decision variables, such as cost minimization or profit maximization.

Step 3 Identify and construct the constraints. Constraints are the linear equations or inequations of the decision variables.

Step 4 Impose the non-negativity restriction on the decision variables.

Some standard LPP formulations are as follows:

5.3.1 Product Mix Problem

An LPP in which there is a competition for limited resources among several products is called a product mix problem.

Statement: Let n products, say, $P_1, P_2, \ldots, P_n$ are to be produced by using $R_1, R_2, \ldots, R_m$ resources that are available in limited amounts of $b_1, b_2, \ldots, b_m$ units respectively. Further let a_{ij}; $i = 1, 2, \ldots, m$; $j = 1, 2, \ldots, n$ denote the number of units of resource R_i required to produce one unit of product P_j and c_j; $j = 1, 2, \ldots, n$ denote the per unit return from the finished product P_j. The problem is to determine the optimum product mix (number of units of each product) that maximize the total over all return under limited resources.

Formulation: Let x_j; $j = 1, 2, \ldots, n$ denote the number of units of the jth product P_j produced.

The above information can be summarized in the form of a table as given below:

		Products					Availability
		P_1	P_2	$\ldots$	P_j	$\ldots$ P_n	
	R_1	a_{11}	a_{12}	$\ldots$	a_{1j}	$\ldots$ a_{1n}	b_1
	$\vdots$	$\vdots$	$\vdots$		$\vdots$	$\vdots$	$\vdots$
Resource	R_i	a_{i1}	a_{i2}	$\ldots$	a_{ij}	$\ldots$ a_{in}	b_i
	$\vdots$	$\vdots$	$\vdots$		$\vdots$	$\vdots$	$\vdots$
	R_m	a_{m1}	a_{m2}	$\ldots$	a_{mj}	$\ldots$ a_{mn}	b_m
Return		c_1	c_2	$\ldots$	c_j	$\ldots$ c_n	

The objective is to maximize the total over all return:

$$Z = c_1 x_1 + c_2 x_2 + \ldots + c_j x_j + \ldots + c_n x_n$$
$$= \sum_{j=1}^{n} c_j x_j \tag{5.4}$$

The total utilization of the ith resource R_i; $i = 1, 2, \ldots, m$ for the proposed production is given by

$$a_{i1} x_1 + a_{i2} x_2 + \ldots + a_{ij} x_j + \ldots + a_{in} x_n = \sum_{j=1}^{n} a_{ij} x_j \text{ units,}$$

which can not exceed the availability b_i of resource R_i. Thus we must have

$$\sum_{j=1}^{n} a_{ij} x_j \leq b_i; \quad i = 1, 2, \ldots, m \tag{5.5}$$

as the ith constraint.

Furthermore, negative units of the products can not be produced which gives the non-negativity restrictions as:

$$x_j \geq 0; \quad j = 1, 2, \ldots, n \tag{5.6}$$

Expressions (5.4), (5.5) & (5.6) give the LPP as:

$$\left. \begin{array}{l} \max\ Z = \sum_{j=1}^{n} c_j x_j \\ \text{subject to } \sum_{j=1}^{n} a_{ij} x_j \leq b_i;\ i = 1, 2, \ldots, m \\ \text{and } x_j \geq 0;\ j = 1, 2, \ldots, n \end{array} \right\} \quad (5.7)$$

5.3.2 Resource Allocation Problem

An LPP which deals with the optimum allocation of limited resources to achieve a given objective is called a resource allocation problem.

Note: Statement and formulation is the same as the Product Mix Problem.

5.3.3 Diet Problem

A balanced diet must contain certain quantities of nutrients such as calories, minerals, vitamins and others. In a diet problem one has to determine the lowest cost diet from a given number of foods to satisfy the minimum requirements of the nutrients. Such a diet is called a "Minimum Cost Balanced Diet".

Statement: Let there be n foods $F_1, F_2, \ldots, F_j, \ldots, F_n$ that provide m nutrients $N_1, N_2, \ldots, N_i, \ldots, N_m$. The minimum requirements of the ith nutrient $N_i;\ i = 1, 2, \ldots, m$ for a balanced diet is b_i units. Also, let $a_{ij};\ i = 1, 2, \ldots, m;\ j = 1, 2, \ldots, n$ denote the amount (in units of measurement) of the ith nutrient present in one unit of the jth food F_j and $c_j;\ j = 1, 2, \ldots, n$ denote the per unit cost of the food F_j. The problem is to determine a "Minimum Cost Balanced Diet".

Formulation: The given information can be summarized as below:

		Foods					
		F_1	F_2	...	F_j	...	F_n	Requirements
	N_1	a_{11}	a_{12}	...	a_{1j}	...	a_{1n}	b_1
	:	:	:	:	:	:	:	:
Nutrients	N_i	a_{i1}	a_{i2}	...	a_{ij}	...	a_{in}	b_i
	:	:	:	:	:	:	:	:
	N_m	a_{m1}	a_{m2}	...	a_{mj}	...	a_{mn}	b_m
Cost/unit (in $)		c_1	c_2	...	c_j	...	c_n	

The objective is to minimize the total cost of the balanced diet.

Let x_j denote the number of units of jth food F_j purchased. The total cost is then given by

$$C = c_1 x_1 + c_2 x_2 + \ldots + c_j x_j + \ldots + c_n x_n$$
$$= \sum_{j=1}^{n} c_j x_j \tag{5.8}$$

The total availability of the ith nutrient N_i; $i = 1, 2, \ldots, m$ from all the n foods purchased is given by

$$a_{i1} x_1 + a_{i2} x_2 + \ldots + a_{ij} x_j + \ldots + a_{in} x_n = \sum_{j=1}^{n} a_{ij} x_j \text{ units,}$$

which must be greater than or equal to the requirement of the ith nutrient N_i. This gives the constraints

$$\sum_{j=1}^{n} a_{ij} x_j \geq b_i; \quad i = 1, 2, \ldots, m \tag{5.9}$$

as the ith constraint.

Furthermore, negative units of the foods can not be purchased, this gives the non-negativity restrictions as

$$x_j \geq 0; \quad j = 1, 2, \ldots, n \tag{5.10}$$

Combining Eqns. (5.8), (5.9) & (5.10) the problem of obtaining a minimum cost balanced diet can be give as the LPP:

$$\left. \begin{aligned} \min \; C &= \sum_{j=1}^{n} c_j x_j \\ \text{subject to} \; &\sum_{j=1}^{n} a_{ij} x_j \geq b_i; \; i = 1, 2, \ldots, m \\ \text{and} \; &x_j \geq 0; \; j = 1, 2, \ldots, n \end{aligned} \right\} \tag{5.11}$$

5.3.4 The Capital Budgeting Problem

A capital budgeting problem is concerned with the selection of the projects available for investment to an investment firm/individual to maximize the returns over a specified period.

Statement: An amount D is available for investment to an investment agency. In the market n projects $P_1, P_2, \ldots, P_j, \ldots, P_n$ are available for investment. Let $a_j;\ j = 1, 2, \ldots, n$ denote the amount required to under take the project P_j and let p_j denote the present worth of all future returns from the project P_j.

The problem is to select the projects for investment that maximize the total returns without exceeding the available budget D.

Formulation: Let

$$x_j = \begin{cases} 1, & \text{if } j^{\text{th}} \text{ project is selected} \\ 0, & \text{if } j^{\text{th}} \text{ project is not selected} \end{cases} \qquad (5.12)$$

The total present worth of all future returns from the selected projects is given as

$$Z = p_1 x_1 + p_2 x_2 + \ldots + p_j x_j + \ldots + p_n x_n = \sum_{j=1}^{n} p_j x_j \qquad (5.13)$$

that is to be maximized.

The total investment in the selected projects is $(a_1 x_1 + a_2 x_2 + \ldots + a_j x_j + \ldots + a_n x_n)$ which can not be greater than the available amount D. This gives the single constraint

$$a_1 x_1 + a_2 x_2 + \ldots + a_j x_j + \ldots + a_n x_n = \sum_{j=1}^{n} a_j x_j \leq D \qquad (5.14)$$

Combining Eqns. (5.12) (5.13) and (5.14), we may write the capital budgeting problem as the following Zero-One LPP:

$$\left. \begin{array}{c} \max\ Z = \sum_{j=1}^{n} p_j x_j \\ \text{subject to } \sum_{j=1}^{n} a_{ij} x_j \leq D \\ \text{and } x_j = 0 \text{ or } 1;\ j = 1, 2, \ldots, n \end{array} \right\} \qquad (5.15)$$

5.3.5 Assignment Problem

Assignment problem is one of the typical integer programming applications in which n workers or machines may be assigned to n jobs. The sole objective might be minimizing the assignment cost of worker i to job j or task completion time of machine i to job j. The mathematical LP model can be formulated as follows. Let

$$x_{ij} = \begin{cases} 1, & \text{if machine i is assigned to job j,} \\ 0, & \text{otherwise} \end{cases} \qquad (5.16)$$

and t_{ij} be the time to complete job j by machine i, then the complete LP model is given by:

$$\left.\begin{array}{l} \min \ Z = \sum_{i=1}^{n}\sum_{j=1}^{n} t_{ij}x_{ij} \\ \text{subject to } \sum_{j=1}^{n} x_{ij} = 1, \ i = 1,2,3,\ldots,n \\ \phantom{\text{subject to }} \sum_{i=1}^{n} x_{ij} = 1, \ j = 1,2,3,\ldots,n \\ \phantom{\text{subject to }} x_{ij} = 0 \ or \ 1 \end{array}\right\} \quad (5.17)$$

5.3.6 Transportation Problem

Transportation problem arises when goods or services are to be transported from one location (source i) to another (destination j). Let x_{ij} be the amount of goods transported from the supply origin or source i to demand area or destination j, and let c_{ij} be the unit cost of this transportation, with a_i and b_j being the available amount at source i and demand at destination j respectively. The objective is to minimize the total cost of transportation. Mathematically, the LP model can be written as

$$\left.\begin{array}{l} \min \ Z = \sum_{i=1}^{m}\sum_{j=1}^{n} c_{ij}x_{ij} \\ \text{subject to } \sum_{j=1}^{n} x_{ij} = a_i, \ i = 1,2,3,\ldots,m \\ \phantom{\text{subject to }} \sum_{i=1}^{m} x_{ij} = b_j, \ j = 1,2,3,\ldots,n \\ \phantom{\text{subject to }} x_{ij} \geq 0 \end{array}\right\} \quad (5.18)$$

5.4 Integer Programming

In LPP, decision variables can take any non-negative value for a particular problem within its feasible region. However, it is called an all integer programming if all the decision variables are constrained to take only integer values. In contrast, it is referred to as a zero-one programming problem when all the decision variables are allowed to take a value of either zero or one, as in the case of assignment problems. Mathematically, the LPP can be formulated as:

$$\left.\begin{array}{l} \min\ (\max)\ \sum_{j=1}^{n} c_j x_j \\ \text{subject to}\ \sum_{j=1}^{n} a_{ij} x_j \leq b_i, for\ i = 1, 2, \ldots, m \\ x_j \geq 0,\ for\ j = 1, 2, \ldots, n. \end{array}\right\} \quad (5.19)$$

In light of Eqn. (5.19), if an additional constraint states that, all x_j assume integer values, it is *pure-integer programming*, and if some values of x_j are integer and others are continuous then, it is *mixed-integer programming*. At the same time, if all the values of x_j are zeros and ones, then the problem becomes *zero-one integer programming*. The areas of integer programming applications include assignment problems, transportation problems, optimum warehouse location and research and development amongst others.

Generally, an integer programming is a particular case of LPP. Various methods for solving the LPP include graphical, simplex, cutting plane, and branch-and-bound (BB) techniques. The graphical method can be used when the problem has only two decision variables, whereas the simplex method is used for more than two variables. The cutting plane techniques are used in obtaining an optimum integer solution, and it is an additional constraint known as *cut* serving as a step in the process. Simultaneously, the BB method is an efficient enumeration technique for obtaining an optimal solution for specific integer programming problems.

In the following sections, some of the standard methods to solve LPP are discussed.

5.5 Graphical Solution of LPP

The LPP of two decision variables can be solved graphically as:

Step 1 Find out the coordinates of each constraint by assuming them as an equality.

Step 2 Plot the constraints on the graph as straight lines.

Step 3 Identify the feasible region. If the origin satisfies the constraint, then the points on the line and towards the origin are feasible for that particular constraint. If the origin does not satisfy the constraint, then points on the line and away from the origin are feasible. Shade the common region that is feasible for all the constraints.

Step 4 Identify the extreme points of the feasible region.

Step 5 Substitute each extreme point in the objective function one by one, and whichever optimizes the objective is the optimum solution of the problem.

Few numerical examples are as follows:

Example 60 Solve the following LP model by using graphical method.

$$\max \ Z = 20x + 10y$$
$$\text{subject to } x + 2y \leq 40$$
$$3x + y \geq 30 \quad (5.20)$$
$$4x + 3y \geq 60$$
$$\text{and } x, y \geq 0$$

Following step 1-4, the graph has been plotted and feasible region identified as shown in Fig. (5.1).

The coordinates of extreme points of the feasible region and the value of the objective function at each point is:

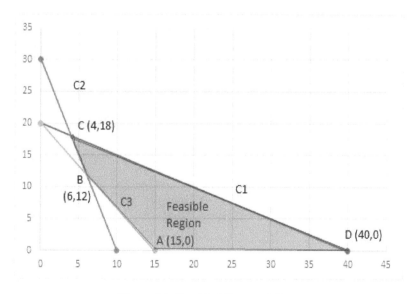

Figure 5.1: Graphical Representation.

Extreme Point	Coordinates (x,y)	Objective Value $Z = 20x + 10y$
A	(15,0)	300
B	(6,12)	240
C	(4,18)	260
D	(40,0)	800

The objective function attains the maximum value at point D, therefore, the optimum solution to the given LP model is $x = 40$, $y = 0$ with Maximum $Z = 800$.

Example 61 *Solve the following LP model by using graphical method.*

$$\max \ Z = 300x + 400y$$
$$\text{subject to } 5x + 4y \leq 200$$
$$3x + 5y \leq 150$$
$$5x + 4y \geq 100$$
$$8x + 4y \geq 80$$
$$\text{and } x, y \geq 0$$

(5.21)

Following step 1-4, the graph has been plotted and feasible region identified as shown in Fig. (5.2). The coordinates of extreme points of the feasible region and the value of the objective function at each point is:

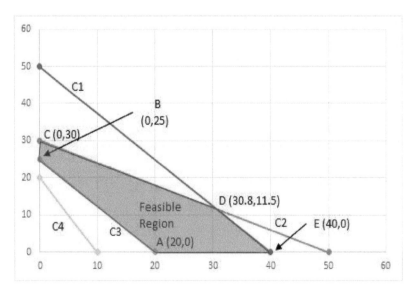

Figure 5.2: Graphical Representation.

Extreme Point	Coordinates (x,y)	Objective Value $Z = 300x + 400y$
A	(20,0)	6,000
B	(0,25)	10,000
C	(0,30)	12,000
D	(30.8,11.5)	13,840
E	(40,0)	12,000

The objective function attains the maximum value at point D, therefore, the optimum solution to the given LP model is $x = 30.8$, $y = 11.5$ with Maximum $Z = 13,840$.

Example 62 Solve the following LP model by using graphical method.

$$\min \ Z = 3x + 2y$$
$$\text{subject to } 5x + y \geq 10$$
$$x + y \geq 6 \qquad (5.22)$$
$$x + 4y \geq 12$$
$$\text{and } x, y \geq 0$$

Following step 1-4, the graph has been plotted and the feasible region identified as shown in Fig. (5.3). The coordinates of the extreme points of the feasible region and the value of objective function at each point is:

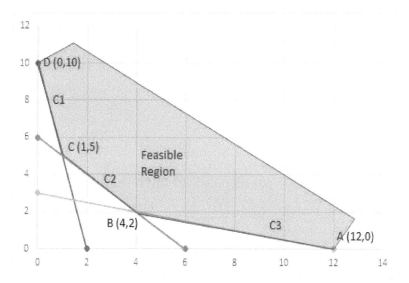

Figure 5.3: Graphical Representation.

Extreme Point	Coordinates (x,y)	Objective Value $Z = 3x+2y$
A	(12,0)	36
B	(4,2)	16
C	(1,5)	13
D	(0,10)	20

The objective function attains the minimum value at point C, therefore, the optimum solution to the given LP model is $x = 1$, $y = 5$ with Minimum $Z = 13$.

Example 63 *Solve the following LP model by using graphical method.*

$$\min \ Z = 200x + 400y$$
$$\text{subject to } x + 3y \geq 400$$
$$x + 2y \leq 350 \quad (5.23)$$
$$\text{and } x, y \geq 0$$

Following step 1-4, the graph has been plotted and feasible region identified as shown in Fig. (5.4). The coordinates of the extreme points of the feasible region and the value of the objective function at each point is:

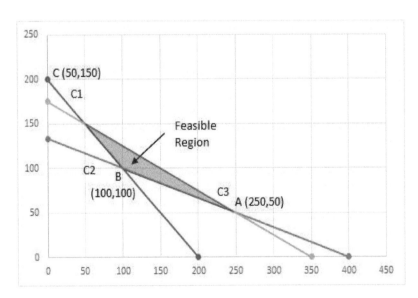

Figure 5.4: Graphical Representation.

The objective function attains the minimum value at point B, therefore, the optimum solution to the given LP model is $x = 100$, $y = 100$ with Minimum $Z = 60,000$.

Extreme Point	Coordinates (x, y)	Objective Value $Z = 200x + 400y$
A	(250, 50)	70,000
B	(100, 100)	60,000
C	(50, 150)	70,000

Remarks:

- When an LP model yields more than one solution which gives same optimum value of objective function, then such a solution is called an alternate (multiple) optimal solution.

- Sometimes, due to the error in problem formulation, there may be infinite solutions to a LPP. Such a solution is known as an unbounded solution.

- When there is no common (or unique) feasible region found, then the LPP has an infeasible solution.

5.6 Theory of Simplex Method

An LPP with two decision variables can be solved by a graphical method, in which linear constraints are plotted on the graph, and the feasible region is identified. Any one extreme point of the feasible region gives the optimal solution to the problem. But, in most real-life problems, more than two decision variables are involved and in that case, the n-dimensional graph cannot be plotted. Therefore, in 1947 G. B. Dantzing developed an iterative method named "Simplex Method" to solve n-dimensional LP models.

The LPP consist of a finite number of feasible solutions. A set of all feasible solutions makes a convex set, and this convex set makes a convex polyhedron, with one extreme point giving the optimal solution to the LPP. The simplex method is also called an "Iterative Method" since it examines each extreme point of the convex feasible region by iteration.

5.6.1 Standard Form

For the use of a simplex method, the given LPP must be converted into standard form. A standard form consists of three characteristics:

- By adding slack[1]/surplus[2]/artificial variables all the constraints should be written as an equation.

---
[1] An unused resource is called a Slack variable. These variables do not make a profit; therefore such variables are added to the objective function with 0 coefficients.
[2] A variable represented exceeding the amount of resource is called a Surplus variable. Likewise, the slack variables are also added with 0 coefficient in the objective function.

- All the right-hand side parameters should be positive, and if not, then made positive by multiplying both sides of the constraint by (-1).

- The objective function should be of maximization type, if not, then convert the minimization problem into equivalent maximization problem as

$$\min \sum_{j=1}^{n} c_j x_j = -\max \sum_{j=1}^{n} (-c_j) x_j$$

Consider the LPP in the standard form:

$$\max \ Z = \sum_{j=1}^{n} c_j x_j + \sum_{i=1}^{m} 0 s_i$$

$$\text{subject to } \sum_{j=1}^{n} a_{ij} x_j + s_i = b_i; \ i = 1, 2, \ldots, m \tag{5.24}$$

$$\text{and } x_j, s_i \geq 0 \ \forall \ i \ \& \ j$$

The set of feasible solution to the LPP (5.24) is given by

$$F = \{(x_1, x_2, \ldots, x_n) | \sum_{j=1}^{n} a_{ij} x_j + s_i = b_i \text{ and } x_j, s_i \geq 0 \ \forall \ i \ \& \ j\}$$

Assumptions:

(i) $F \neq \phi \Rightarrow$ there exist at least one feasible solution to the LPP (5.24).

(ii) All basic feasible solutions to the LPP (5.24) are non-degenerate, which implies that there are exactly m positive x_j in every basic feasible solution.

(iii) A basic feasible solution $\underline{x}_0$ is known. Let it be $\underline{x}_0 = (x_{10}, x_{20}, \ldots, x_{mo}, 0, \ldots, 0)$. Without loss of generality it could be assumed that the first m components of $\underline{x}_0$ are positive and last (n-m) components are zero. By the property of basic feasible solutions, the vectors a_{ij} associated with positive $x'_j s$ will be linearly independent.

5.6.2 Computational Procedure of Simplex Method

Following the given steps an LPP can be solved by simplex method.

Step 1 Convert the given LPP in the standard form

$$\max \ Z = \underline{c}' \underline{x}, \ s.t. \ A\underline{x} + s = b \ \& \ \underline{x}, s \geq 0$$

where A is an $m \times n$ matrix of known real numbers.

Step 2 Select any m columns of A to form a square sub-matrix B such that $|B| \neq 0$.

Step 3 Compute B^{-1} and then $B^{-1}\underline{b}$. If $B^{-1}\underline{b} \not\geq \underline{0}$, discard this B and go back to step 2, otherwise go to step 4 with $B^{-1}\underline{b} \geq \underline{0}$.

Step 4 Compute $\underline{c}'_B B^{-1}\underline{b}$, $B^{-1}A$ and $z_j - c_j = \underline{c}'_B B^{-1}\underline{a}_j - c_j$, where $\underline{c}'_B$ is the vector of the coefficients of the basic variables in the objective function $Z = \underline{c}'\underline{x}$. Set up the starting simplex tableau as shown in Table (5.1):

Table 5.1: Tableau '0'.

$\underline{x}_B$	$\underline{c}_B$	c_j RHS	c_1 x_1	$\cdots$	$\cdots$	c_n x_n	0 s_1	$\cdots$	0 s_m
x_{B_1}	c_{B_1}		a_{11}	$\cdots$	$\cdots$	a_{1n}	1	$\cdots$	0
$\vdots$	$\vdots$		$\vdots$	$\cdots$	$\cdots$	$\vdots$	$\vdots$	$\cdots$	$\vdots$
x_{B_r}	c_{B_r}	$B^{-1}\underline{b}$	a_{r1}	$\cdots$	$\cdots$	a_{rn}	0	1	0
$\vdots$	$\vdots$	$= \overline{\underline{b}}$	$\vdots$	$\cdots$	$\cdots$	$\vdots$	$\vdots$	$\cdots$	$\vdots$
x_{B_m}	c_{B_m}		a_{m1}	$\cdots$	$\cdots$	a_{mn}	0	$\cdots$	1
		$z_j = \underline{c}'_B B^{-1}\underline{b}$	0	$\cdots$	$\cdots$	0	0	$\cdots$	0
		$z_j - c_j$	$z_1 - c_1$	$\cdots$	$\cdots$	$z_n - c_n$	0	$\cdots$	0

where $\underline{x}_B$ is the vector of basic variables.

Step 5 For the optimality test, if all the $z_j - c_j \geq 0$, STOP, the current solution is optimum. Otherwise, go to step 6.

Step 6 Let $\underline{y}_k$ denote the kth column of $B^{-1}A$, then $\underline{y}_k = B^{-1}\underline{a}_k$. If $\underline{y}_k \leq \underline{0}$, STOP, the given LPP has an unbounded solution. Otherwise go to step 7.

Step 7 Determine x_{B_r} as follows:

$$\frac{\overline{b}_r}{y_{rk}} = \min_i \left\{ \frac{\overline{b}_i}{y_{ik}} | y_{ik} > 0 \right\}$$

where $\overline{b}_r$ is the rth component of $B^{-1}\underline{b} = \overline{\underline{b}}$. In the improved solution x_{B_r} will become non-basic and x_r will become basic in its place. Update the simplex tableau and go to step 5. Repeat steps 5 to 7 until a 'STOP' is reached.

Notes:

(1) If no B with $|B| \neq 0$ and $B^{-1}\underline{b} \geq \underline{0}$, is found (see steps 2 & 3) STOP, the given LPP has no solution.

(2) When at step 2 $B = I_m$ the computations to set up Tableau '0' simplifies to a great extent because: $B^{-1} = I_m$ and if $B^{-1}\underline{b} = \underline{b} \geq 0$, we have $\underline{c}'_B B^{-1}\underline{b} = \underline{c}'_B \underline{b}$, $B^{-1}A = A$, $z_j - c_j = \underline{c}'_B \underline{a}_j - c_j$. $Z = \underline{c}'_B B^{-1}\underline{b} = \underline{c}'_B \underline{b}$ and $\underline{y}_k = B^{-1}\underline{a}_k = \underline{a}_k$. Furthermore, if at the starting stage all the basic variables are slack or surplus variables we have $\underline{c}_B = \underline{0}$ and the computations are further simplified as:

$$\underline{c}'_B B^{-1}\underline{b} = 0, \; z_j - c_j = \underline{c}'_B B^{-1}\underline{a}_j - c_j = -c_j$$

and $Z = \underline{c}'_B B^{-1}b = 0$.

(3) **Updating Basis:** To form the new basis one variable enters the basis and one leaves the basis with the associated value under c_B column. Convert the key element to unit by dividing the key row by key element and rest of the other elements of key column to zero by using the formula:

$$\text{New element} = \text{old element} - \left\{ \frac{\text{Product of elements in key row and column}}{\text{key element}} \right\}$$

5.6.3 Maximization Problem

Example 64 *Use simplex method to solve the LPP.*

$$\begin{aligned} \max \; Z &= 3x_1 + 2x_2 \\ \text{subject to } x_1 + x_2 &\leq 4 \\ x_1 - x_2 &\leq 2 \\ \text{and } x_1, x_2 &\geq 0 \end{aligned} \tag{5.25}$$

Solution Step 1: Standard Form—To convert the inequality ($\leq$) constraints to equality, slack variables s_1 and s_2 are added in each constraint. Then, the standard form will be:

$$\left. \begin{aligned} \max \; Z &= 3x_1 + 2x_2 + 0s_1 + 0s_2 \\ \text{subject to } x_1 + x_2 + s_1 &= 4 \\ x_1 - x_2 + s_2 &= 2 \\ \text{and } x_1, x_2, s_1, s_2 &\geq 0 \end{aligned} \right\} \tag{5.26}$$

Step 2: The initial basic feasible solution is corresponding to s_1 and s_2, therefore

$$B = \begin{bmatrix} 1 & 0 \\ 0 & 1 \end{bmatrix}$$

Step 3:

$$B^{-1} = \begin{bmatrix} 1 & 0 \\ 0 & 1 \end{bmatrix}$$

$$B^{-1}\underline{b} = \begin{bmatrix} 1 & 0 \\ 0 & 1 \end{bmatrix} \begin{bmatrix} 4 \\ 2 \end{bmatrix} = \begin{bmatrix} 4 \\ 2 \end{bmatrix}$$

Step 4: The starting simplex tableau is as follows:

Table 5.2: Tableau '0'.

		c_j	3	2	0	0
c_B	Basis	x_B	x_1	x_2	s_1	s_2
0	s_1	4	1	1	1	0
←0	s_2	2	1	-1	0	1
		z_j	0	0	0	0
		$z_j - c_j$	-3↑	-2	0	0

Since all $z_j - c_j \not\geq 0$, therefore the current basic feasible solution given in Table (5.2) is not optimal and we have to update the solution. Now x_1 corresponding to the most negative value of $z_j - c_j$ will enter the basis. To identify the leaving variable minimum ratio between x_B and key column has been calculated and it is found that s_2 will leave the basis as the minimum ratio $2/1 = 2$ is corresponding to s_2.

To update the solution convert the key element, i.e., 1 into unity by dividing the key row by key element and rest of the other elements in key column to zero by using the formula given below:

$$New\ element = old\ element - \left\{\frac{Product\ of\ elements\ in\ key\ row\ and\ column}{key\ element}\right\}$$

In tableau 0, the key element is already unity, therefore the key row will be same in the tableau 1. Row one will be changed in conversion of the other element of the key column to zero as:

$$New\ element = 4 - \left\{\frac{1 \times 2}{1} = 2\right\}$$

Similarly, calculations for other elements can be done.

Since all $z_j - c_j \not\geq 0$, therefore the current basic feasible solution given in Table (5.3) is not optimal and we have to again update the solution following the same procedure. Now the entering variable is x_2 and leaving variable is s_1. The modified solution is given in Table (5.4):

Since all $z_j - c_j \geq 0$, the current solution is optimum. The optimal solution is Max $Z = 11$, $x_1 = 3$ and $x_2 = 1$. ∎

Table 5.3: Tableau '1'.

		c_j	3	2	0	0
c_B	Basis	x_B	x_1	x_2	s_1	s_2
←0	s_1	2	0	2	1	-1
3	x_1	2	1	-1	0	1
		z_j	3	-3	0	0
		$z_j - c_j$	0	-5↑	0	0

Table 5.4: Tableau '2'.

		c_j	3	2	0	0
c_B	Basis	x_B	x_1	x_2	s_1	s_2
2	x_2	1	0	1	1/2	-1/2
3	x_1	3	1	0	1/2	1/2
		z_j	3	2	5/2	1/2
		$z_j - c_j$	0	0	5/2	1/2

5.6.4 Degeneracy and Cycling in LPP

A feasible solution of an LPP is degenerate if one or more of the basic variables vanish. Let vector $(P_1, P_2, \ldots, P_n)$ form a degenerate basis then in the constraint equation $x_1 P_1 + x_2 P_2 + \cdots + x_n P_n = P_0$ at least one $x_j = 0$.

In the theory of the simplex method, it was assumed that all basic feasible solutions are non-degenerate. This assumption is necessary for improving the value of the objective function at each successive iteration for obtaining the optimal solution in a finite number of iterations. In the case of degeneracy $\theta_0 = 0$, there may be no improvement in the value of the objective function at two consecutive iterations. It is then possible to have the case that the same sequence of basis is repeatedly selected without ever satisfying the optimality condition and this is referred to as "Cycling".

Degeneracy may occur if $\theta_0 = \min_i \{ \frac{x_{i0}}{x_{ik}} | x_{ik} > 0 \}$ is not attained for a unique i. The next step solution will be a degenerate, and it is possible to have $\theta_0 = 0$ in the next iteration and hence cycling may occur. No practical LPP has cycling problems; it is only for theoretical discussion.

5.6.5 Artificial Basis Technique

When no identity basis is formed, artificial variables are used for making an identity basis to start the simplex method procedure. In the LPP having minimization type objective function, artificial variables are added with the coefficient "+M," where M is a large positive number that is not specified. Whereas in maximization type LPP, the artificial variables' coefficients in the objective function will

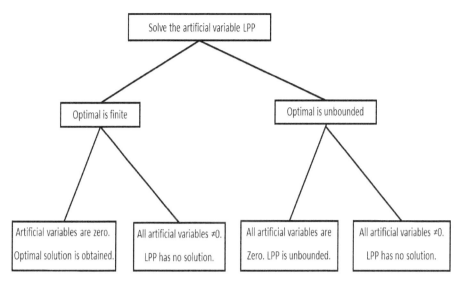

Figure 5.5: Cases Involved in the Artificial Variable LPP.

be "-M." Since M is a large unspecified number hence the method is also known as "Big M method".

The above discussed LPP is treated like a usual LPP, and the ordinary simplex method steps are used to solve it.

The following possible cases are depicted in the Fig. (5.5) which may occur in solving the artificial variable LPP.

5.6.6 Minimization Problem

Example 65 *Use the penalty method to solve the LPP.*

$$\min \ Z = 5y_1 + 3y_2$$
$$\text{subject to } 2y_1 + 4y_2 \leq 12$$
$$2y_1 + 2y_2 = 10 \qquad (5.27)$$
$$5y_1 + 2y_2 \geq 10$$
$$\text{and } y_1, y_2 \geq 0$$

Solution Step 1: Standard Form- Add slack variables s_1 in the first constraint, artificial variables A_1 & A_2 in the second and third constraints, and subtract s_2 in third constraint to make inequality ($\leq, \geq$) as equality constraints. Then, the standard form of LPP is:

$$\min Z = 5y_1 + 3y_2 + 0s_1 + 0s_2 + MA_1 + MA_2$$
$$\text{subject to } 2y_1 + 4y_2 + s_1 \leq 12$$
$$2y_1 + 2y_2 + A_1 = 10 \qquad (5.28)$$
$$5y_1 + 2y_2 - s_2 + A_2 \geq 10$$
$$\text{and } y_1, y_2, s_1, s_2, A_1, A_2 \geq 0$$

Step 2: The initial basic feasible solution is corresponding to s_1, A_1 and A_2, therefore

$$B = \begin{bmatrix} 1 & 0 & 0 \\ 0 & 1 & 0 \\ 0 & 0 & 1 \end{bmatrix}$$

Step 3:

$$B^{-1} = \begin{bmatrix} 1 & 0 & 0 \\ 0 & 1 & 0 \\ 0 & 0 & 1 \end{bmatrix}$$

$$B^{-1}\underline{b} = \begin{bmatrix} 1 & 0 & 0 \\ 0 & 1 & 0 \\ 0 & 0 & 1 \end{bmatrix} \begin{bmatrix} 12 \\ 10 \\ 10 \end{bmatrix} = \begin{bmatrix} 12 \\ 10 \\ 10 \end{bmatrix}$$

Step 4: The initial simplex table is as follows:

Table 5.5: Tableau '0'.

c_B	Basis	c_j x_B	5 x_1	3 x_2	0 s_1	0 s_2	M A_1	M A_2
0	s_1	12	2	4	1	0	0	0
M	A_1	10	2	2	0	0	1	0
←M	A_2	10	5	2	0	-1	0	1
		z_j	7M	4M	0	-M	M	M
		$z_j - c_j$	7M-5↑	4M-3	0	-M	0	0

In case of minimization, $z_j - c_j \leq 0$ for optimality and since all $z_j - c_j \nleq 0$, therefore the current basic feasible solution given in Table (5.5) is not optimal and we have to update the solution. Now x_1 corresponding to the most positive value of $z_j - c_j$ will enter the basis. To identify the leaving variable minimum ratio between x_B and key column has been calculated and it is found that A_2 will leave the basis as the minimum ratio $10/5 = 2$ is corresponding to A_2.

To update the solution convert the key element, i.e., 5 into unity by dividing the key row by 5 and rest of the other elements in the key column to zero by using the formula given below:

$$\text{New element} = \text{old element} - \left\{ \frac{\text{Product of elements in key row and column}}{\text{key element}} \right\}$$

Row one will be changed in the conversion of the other element of the key column to zero as:

$$New\ element = 12 - \left\{\frac{10 \times 2}{5} = 8\right\}$$

Similarly, calculations for other elements can be done.

Table 5.6: Tableau '1'.

c_B	Basis	c_j x_B	5 x_1	3 x_2	0 s_1	0 s_2	M A_1
←0	s_1	8	0	16/5	1	2/5	0
M	A_1	6	0	6/5	0	2/5	1
5	x_1	2	1	2/5	0	-1/5	0
		z_j	5	6M/5+2	0	2M/5-1	M
		$z_j - c_j$	0	6M/5-1↑	0	2M/5-1	0

Since all $z_j - c_j \not\leq 0$, therefore the current basic feasible solution given in Table (5.6) is not optimal and we have to again update the solution following the same procedure. Now the entering variable is x_2 and leaving variable is s_1.

Table 5.7: Tableau '2'.

c_B	Basis	c_j x_B	5 x_1	3 x_2	0 s_1	0 s_2	M A_1
3	x_2	5/2	0	1	5/16	1/8	0
←M	A_1	3	0	0	-3/8	1/4	1
5	x_1	1	1	0	-1/8	-1/4	0
		z_j	5	3	-3M/8+5/16	M/4-7/8	M
		$z_j - c_j$	0	0	-3M/8+5/16	M/4-7/8↑	0

Since all $z_j - c_j \not\leq 0$, therefore the current basic feasible solution given in Table (5.7) is not optimal and we have to again update the solution following the same procedure. Now the entering variable is s_2 and leaving variable is A_1.

Since all $z_j - c_j \leq 0$, the current solution is optimum. The optimal solution is Min Z = 23, $x_1 = 4, x_2 = 1, s_1 = 0$ and $s_2 = 12$. ∎

5.7 Solving LPP using LINGO-18

In the previous sections, the LP models have been solved either by graphical method (when two variables are involved) or through the simplex computational procedure. But, this book focuses on the solution of optimization problems by

Table 5.8: Tableau '3'.

c_B	Basis	c_j x_B	5 x_1	3 x_2	0 s_1	0 s_2
3	x_2	1	0	1	1/2	0
0	s_2	12	0	0	-3/2	1
5	x_1	4	1	0	-1/2	0
		z_j	5	3	-1	0
		$z_j - c_j$	0	0	-1	0

LINGO-18 software. Therefore, some numerical examples of the solution of LPPs through LINGO-18 software are as follows:

5.7.1 Product Mix Profit Maximization Problem

$$\max \ Z = 1.20x_1 + 1.40x_2 \quad \text{(profit)}$$
subject to $40x_1 + 25x_2 \leq 1000$ (Machining capacity)
$$35x_1 + 28x_2 \leq 980 \quad \text{(Boring capacity)} \quad (5.29)$$
$$25x_1 + 35x_2 \leq 875 \quad \text{(Polishing capacity)}$$
and $x_1, x_2 \geq 0$

LINGO-18 codes for the problem (5.29)

MODEL:
SETS:
product/1..2/ : c,x;
capacity/1..3/ : mat;
material(capacity, product) : a;
endsets
!objective function of profit maximization;
[objective]max = @sum(product(j) : c(j)*x(j));
!constraints of machining, boring, and polishing capacities;
@for(capacity(i) : [material_constraints]
@sum(product(j) : a(i, j)*x(j)) <= mat(i));
Data :
c = 1.20, 1.40;
mat = 1000, 980, 875;
a = 40 25 35 28 25 35;
enddata
end

The global optimal solution of LPP (5.29) derived through LINGO-18 software is $x_1 = 16.94$, $x_2 = 12.90$ with maximum profit 38.39.

5.7.2 Product Sales through Advertisement Problem

$$\max \ Z = 400x_1 + 900x_2 + 500x_3 + 200x_4$$
$$\text{subject to } 40x_1 + 75x_2 + 30x_3 + 15x_4 \leq 800$$
$$30x_1 + 40x_2 + 20x_3 + 10x_4 \geq 200$$
$$40x_1 + 75x_2 \leq 500$$
$$x_1 \geq 3 \tag{5.30}$$
$$x_2 \geq 2$$
$$5 \leq x_3 \leq 10$$
$$5 \leq x_4 \leq 10$$
$$\text{and } x_1, x_2, x_3, x_4 \geq 0$$

LINGO-18 codes for the problem (5.30)

```
MODEL:
SETS:
product/1..4/ : c,x;
capacity1/1..4/ : mat;
capacity2/1..5/ : mat1;
material1(capacity1, product) : a;
material2(capacity2, product) : b;
endsets
!objective function of profit maximization;
[objective]max = @sum(product(j) : c(j)*x(j));
!constraints of less than equal to type;
@for(capacity1(i) : [material1_constraints]
@sum(product(j) : a(i,j)*x(j)) <= mat(i));
!constraints of greater than equal to type;
@for(capacity2(i) : [material2_constraints]
@sum(product(j) : b(i,j)*x(j)) >= mat1(i));
Data:
c = 400, 900, 500, 200;
mat = 800, 500, 10, 10;
mat1 = 200, 3, 2, 5, 5;
a = 40 75 30 15  40 75 0 0  0 0 1 0  0 0 0 1;
b = 30 40 20 10  1 0 0 0  0 1 0 0  0 0 1 0  0 0 0 1;
enddata
end
```

The global optimal solution of LPP (5.30) derived through LINGO-18 software is $x_1 = 3$, $x_2 = 3.07$, $x_3 = 10$, $x_4 = 10$ with maximum profit 10960.

5.7.3 Profit Maximization Problem

$$\max \; Z = 13x_1 + 17.5x_2 + 15x_3 + 19.5x_4 + 12x_5 + 19x_6$$

subject to $x_1 + x_2 \leq 200$ (Supply constraints)

$\qquad\qquad x_3 + x_4 \leq 300$

$\qquad\qquad x_5 + x_6 \leq 400$ \hfill (5.31)

$\qquad\qquad x_1 + x_3 + x_5 \leq 450$ (Plant capacities constraints)

$\qquad\qquad x_2 + x_4 + x_6 \leq 550$

and $x_i \geq 0; \; i = 1, 2, \ldots, 6$

LINGO-18 codes for the problem (5.31)

```
MODEL:
SETS:
product/1..6/ : c,x;
supply/1..3/ : mat;
capacity/1..2/ : mat1;
material(supply, product) : a;
plant(capacity, product) : b;
endsets
!objective function of profit maximization;
[objective]max = @sum(product(j) : c(j)*x(j));
!constraints of supply;
@for(supply(i) : [material_constraints]
@sum(product(j) : a(i,j)*x(j)) <= mat(i));
!constraints of plant capacities;
@for(capacity(i) : [plant_constraints]
@sum(product(j) : b(i,j)*x(j)) <= mat1(i));
Data:
c = 13, 17.5, 15, 19.5, 12, 19;
mat = 200, 300, 400;
mat1 = 450, 550;
a = 1 1 0 0 0 0   0 0 1 1 0 0   0 0 0 0 1 1;
b = 1 0 1 0 1 0   0 1 0 1 0 1;
enddata
end
```

The global optimal solution of LPP (5.31) derived through LINGO-18 software is $x_1 = 200$, $x_2 = 0$, $x_3 = 150$, $x_4 = 150$, $x_5 = 0$, $x_6 = 400$ with maximum profit 15375.

5.7.4 Total Production Cost Minimization Problem

$$\min \ Z = 9.7x_1 + 9.7x_2 + 6.9x_3 + 6.9x_4 + 6.3x_5 + 6.8x_6 + 6.8x_7 + 6.8x_8$$

subject to
$$x_1 + x_2 \leq 800$$
$$x_3 + x_4 \leq 700$$
$$x_5 + x_6 \leq 1000$$
$$x_7 + x_8 \leq 600$$
$$x_1 + x_3 + x_5 + x_7 = 1400$$
$$x_2 + x_4 + x_6 + x_8 = 900$$
$$-2x_1 - 0.5x_3 - 3x_5 + 11x_7 \leq 0$$
$$-2x_2 - 0.5x_4 - 3x_6 + 11x_8 \leq 0$$
$$x_1 - 10x_3 - 17x_5 + 3x_7 \geq 0$$
$$-9x_2 - 20x_4 - 27x_6 - 7x_8 \geq 0$$
and $x_i \geq 0; \ i = 1, 2, \ldots, 8$

(5.32)

In the above problem (5.32) first four constraints are material availability constraints, 5th & 6th are product demand, and rest are the product specification constraints.

LINGO-18 codes for the problem (5.32)

```
MODEL:
SETS:
product/1..8/ : c,x;
availability/1..4/ : mat;
demand/1..2/ : mat1;
specification1/1..2/ : mat2;
specification2/1..2/ : mat3;
material(availability, product) : a;
product1(demand, product) : b;
product2(specification1, product) : e;
product3(specification2, product) : d;
endsets
!objective function of cost minimization;
[objective]min = @sum(product(j) : c(j)*x(j));
!constraints of material availability;
@for(availability(i) : [material_constraints]
@sum(product(j) : a(i,j)*x(j)) <= mat(i));
!constraints of product demand;
@for(demand(i) : [product1_constraints]
@sum(product(j) : b(i,j)*x(j)) = mat1(i));
!constraints of product specifications;
@for(specification1(i) : [product2_constraints]
@sum(product(j) : e(i,j)*x(j)) <= mat2(i));
@for(specification2(i) : [product3_constraints]
@sum(product(j) : d(i,j)*x(j)) <= mat3(i));
Data:
c = 9.7, 9.7, 6.9, 6.9, 6.3, 6.8, 6.8, 6.8;
mat = 800, 700, 1000, 600;
mat1 = 1400, 900;
mat2 = 0, 0;
mat3 = 0, 0;
a = 1 1 0 0 0 0 0 0   0 0 1 1 0 0 0 0   0 0 0 0 1 1 0 0   0 0 0 0 0 0 1 1;
b = 1 0 1 0 1 0 1 0   0 1 0 1 0 1 0 1;
e = -2 0 -0.5 0 -3 0 1 1 0  0 -2 0 -0.5 0 -3 0 1 1;
d = 1 0 -1 0 0 -1 7 0 3 0   0 -9 0 -2 0 0 -2 7 0 -7;
enddata
end
```

The global optimal solution of LPP (5.32) derived through LINGO-18 software is $x_1 = 107.69$, $x_2 = 142.31$, $x_3 = 0$, $x_4 = 700$, $x_5 = 1000$, $x_6 = 0$, $x_7 = 292.31$, $x_8 = 57.69$ with minimum total production cost 15935.

5.7.5 Product Manufacturing Problem

$$\max\ Z = 0.40x_1 + 0.28x_2 + 0.32x_3 + 0.72x_4 + 0.64x_5 + 0.60x_6$$

subject to $0.01x_1 + 0.01x_2 + 0.01x_3 + 0.03x_4 + 0.03x_5 + 0.03x_6 \leq 850$

$$0.02x_1 + 0.05x_4 \leq 700$$
$$0.02x_2 + 0.05x_5 \leq 100 \qquad (5.33)$$
$$0.03x_3 + 0.08x_6 \leq 900$$

and $x_i \geq 0;\ i = 1, 2, \ldots, 6$

LINGO-18 codes for the problem (5.33)

$MODEL:$
$SETS:$
$product/1..6/ : c, x;$
$capacity/1..4/ : mat;$
$material(capacity, product) : a;$
$endsets$
!objective function of profit maximization;
$[objective]max = @sum(product(j) : c(j) * x(j));$
!constraints;
$@for(capacity(i) : [material\_constraints]$
$@sum(product(j) : a(i,j) * x(j)) <= mat(i));$
$Data:$
$c = 0.40, 0.28, 0.32, 0.72, 0.64, 0.60;$
$mat = 850, 700, 100, 900;$
$a = 0.01\ 0.01\ 0.01\ 0.03\ 0.03\ 0.03\quad 0.02\ 0\ 0\ 0.05\ 0\ 0$
$0\ 0.02\ 0\ 0\ 0.05\ 0\quad 0\ 0\ 0.03\ 0\ 0\ 0.08;$
$enddata$
end

The global optimal solution of LPP (5.33) derived through LINGO-18 software is $x_1 = 35,000,\ x_2 = 5000,\ x_3 = 30,000,\ x_4 = 0,\ x_5 = 0,\ x_6 = 0$ with maximum profit 25,000.

5.7.6 Product Manufacturing Problem

$$\max\ Z = 5x_1 + 7x_2$$

subject to $2x_1 + 5x_2 \leq 51$ (Material constraints)

$3x_1 + 2x_2 \leq 42$ (Man Hour Constraint)

$x_1 + x_2 \geq 14$ (Total No. of Products) $\qquad (5.34)$

$x_2 \geq 7$ (Plant capacities constraints)

and $x_1, x_2 \geq 0$

> **LINGO-18 codes for the problem (5.34)**
>
> MODEL :
> SETS :
> $product/1..2/:c,x;$
> $const1/1..2/:mat;$
> $const2/1..2/:mat1;$
> $material1(const1,product):a;$
> $material2(const2,product):b;$
> endsets
> !objective function of profit maximization;
> $[objective]max = @sum(product(j):c(j)*x(j));$
> !constraints;
> $@for(const1(i)):[material1\_constraints]$
> $@sum(product(j):a(i,j)*x(j))<=mat(i));$
> $@for(const2(i)):[material2\_constraints]$
> $@sum(product(j):b(i,j)*x(j))>=mat1(i));$
> Data :
> $c = 5,7;$
> $mat = 51,42;$
> $mat1 = 14,7;$
> $a = 2\ 5\ \ 3\ 2;$
> $b = 1\ 1\ \ 0\ 1;$
> enddata
> end

The global optimal solution of LPP (5.34) derived through LINGO-18 software is $x_1 = 8$, $x_2 = 7$ with maximum profit 89.

5.7.7 Production Cost Minimization Problem

$$\min\ Z = 0.25x_1 + 0.17x_2 + 0.18x_3 + 0.25x_4 + 0.18x_5 + 0.14x_6$$

subject to

Plant I : $x_1 + x_2 + x_3 \leq 850$ (Production Capacity constraints)

Plant II : $x_4 + x_5 + x_6 \leq 650$

Warehouse A : $x_1 + x_4 = 300$ (Total Demand constraints)

Warehouse B : $x_2 + x_5 = 400$

Warehouse C : $x_3 + x_6 = 500$

and $x_i \geq 0;\ i = 1, 2, \ldots, 6$

(5.35)

LINGO-18 codes for the problem (5.35)

$MODEL$:
$SETS$:
$product/1..6/ : c,x;$
$capacity/1..2/ : mat;$
$demand/1..3/ : mat1;$
$plant(capacity, product) : a;$
$warehouse(demand, product) : b;$
$endsets$
!objective function of cost minimization;
$[objective]min = @sum(product(j) : c(j) * x(j));$
!constraints of production capacity;
$@for(capacity(i) : [plant\_constraints]$
$@sum(product(j) : a(i,j) * x(j)) <= mat(i));$
!constraintsoftotaldemand;
$@for(demand(i) : [warehouse\_constraints]$
$@sum(product(j) : b(i,j) * x(j)) = mat1(i));$
$Data$:
$c = 0.25, 0.17, 0.18, 0.25, 0.18, 0.14;$
$mat = 850, 650;$
$mat1 = 300, 400, 500;$
$a = 1\ 1\ 1\ 0\ 0\ 0\ \ \ 0\ 0\ 0\ 1\ 1\ 1;$
$b = 1\ 0\ 0\ 1\ 0\ 0\ \ \ 0\ 1\ 0\ 0\ 1\ 0\ \ \ 0\ 0\ 1\ 0\ 0\ 1;$
$enddata$
end

The global optimal solution of LPP (5.35) derived through LINGO-18 software is $x_1 = 300$, $x_2 = 400$, $x_3 = 0$, $x_4 = 0$, $x_5 = 0$, $x_6 = 500$ with minimum cost 213.

5.7.8 Profit Maximization Problem

$$\max \ Z = 5x_{11} + 4x_{12} + 4x_{13} + 3x_{14} + 6x_{21} + 2x_{22} + 3x_{23} + 4x_{24} + 10x_{31} + 5x_{32} + 6x_{33} + 2x_{34}$$

subject to $x_{11} + x_{12} + x_{13} + x_{14} \leq 35$ (Supply constraints)

$$x_{21} + x_{22} + x_{23} + x_{24} \leq 50$$
$$x_{31} + x_{32} + x_{33} + x_{34} \leq 40$$
$$x_{11} + x_{21} + x_{31} \geq 45 \quad \text{(Demand constraints)}$$
$$x_{12} + x_{22} + x_{32} \geq 20$$
$$x_{13} + x_{23} + x_{33} \geq 30$$
$$x_{14} + x_{24} + x_{34} \geq 30$$

and $x_{ij} \geq 0 \ \forall \ i = 1,2,3; \ j = 1,2,3,4$

(5.36)

LINGO-18 codes for the problem (5.36)

MODEL:
!Job Allocation Problem;
SETS:
$Job/123/ : mat$;
$Machine/1234/ : mat1$;
$Links(Job, Machine) : Profit, Assign$;
endsets
!Objective;
$max = @sum(Links(i,j) : Profit(i,j) * Assign(i,j))$;
!Constraints;
$@for(Job(i) : @sum(Machine(j) : Assign(i,j)) <= mat(i))$;
$@for(Machine(j) : @sum(Job(i) : Assign(i,j)) >= mat1(j))$;
Data:
Profit = 5 4 4 3 6 2 3 4 10 5 6 2;
mat = 35, 50, 40;
mat1 = 45, 20, 30, 30;
enddata
end

The global optimal solution of LPP (5.36) derived through LINGO-18 software is $x_{11} = 0$, $x_{12} = 20$, $x_{13} = 15$, $x_{14} = 0$, $x_{21} = 5$, $x_{22} = 0$, $x_{23} = 15$, $x_{24} = 30$, $x_{31} = 40$, $x_{32} = 0$, $x_{33} = 0$, $x_{34} = 0$ with maximum profit 735.

5.7.9 Cost Minimization Problem

min $Z = 8x_{11} + 6x_{12} + 10x_{13} + 9x_{14} + 9x_{21} + 12x_{22} + 13x_{23} + 7x_{24} + 14x_{31} + 9x_{32} + 16x_{33} + 5x_{34}$

subject to $x_{11} + x_{12} + x_{13} + x_{14} \leq 35$ (Supply constraints)

$x_{21} + x_{22} + x_{23} + x_{24} \leq 50$

$x_{31} + x_{32} + x_{33} + x_{34} \leq 40$

$x_{11} + x_{21} + x_{31} \geq 45$ (Demand constraints)

$x_{12} + x_{22} + x_{32} \geq 20$

$x_{13} + x_{23} + x_{33} \geq 30$

$x_{14} + x_{24} + x_{34} \geq 30$

and $x_{ij} \geq 0 \ \forall \ i = 1,2,3; \ j = 1,2,3,4$

(5.37)

LINGO-18 codes for the problem (5.37)

$MODEL:$
!Job Allocation Problem;
$SETS:$
$Job/123/:mat;$
$Machine/1234/:mat1;$
$Links(Job, Machine): Profit, Assign;$
$endsets$
!Objective;
$max = @sum(Links(i,j): Profit(i,j) * Assign(i,j));$
!Constraints;
$@for(Job(i): @sum(Machine(j): Assign(i,j)) <= mat(i));$
$@for(Machine(j): @sum(Job(i): Assign(i,j)) >= mat1(j));$
$Data:$
$Profit = 8\ 6\ 10\ 9 \quad 9\ 12\ 13\ 7 \quad 14\ 9\ 16\ 5;$
$mat = 35, 50, 40;$
$mat1 = 45, 20, 30, 30;$
$enddata$
end

The global optimal solution of LPP (5.37) derived through LINGO-18 software is $x_{11} = 0$, $x_{12} = 10$, $x_{13} = 25$, $x_{14} = 0$, $x_{21} = 45$, $x_{22} = 0$, $x_{23} = 5$, $x_{24} = 0$, $x_{31} = 0$, $x_{32} = 10$, $x_{33} = 0$, $x_{34} = 30$ with minimum cost 1020.

5.8 Concept of Duality in LPP

Lets consider the standard form of a LPP as follows:

$$\left.\begin{array}{l} \max\ Z = CX \\ \text{subject to } AX = B \\ \text{and } X \geq 0 \end{array}\right\} \quad (5.38)$$

Here, A represents a $m \times n$ matrix such that m is the number of equations and n the number of variables. B is an $m \times 1$ right-hand side column vector, C is the coefficient associated with the decision variables in the objective function, and X the vector of decision variables. The system in Eqn. (5.38) is called primal. Associated with any primal problem there exists an equivalent dual problem with variables $y_1, y_2, \ldots, y_m$. The dual variables are defined relative to the given primal basis, and hence, a change in the basic variables changes the dual values. The primal's initial basic variables are obtained by setting $(n-m)$, the non-basic variables to zero and solving for the remaining m basic variables. Multiplying the primal constraints by their corresponding dual variables and summing over the product then subtracting it from the objective function Z, automatically eliminates the m basic variables.

Let $Y^T = (y_1, y_2, \ldots, y_m)$. Multiplying Y^T by the constraints in Eqn. (5.38) and subtracting from Z, gives

$$Z - Y^T B = (C - Y^T A)X = \bar{C}X \quad (5.39)$$

where $\bar{C}X = C - Y^T A$ is the relative cost factor. If $y_1, y_2, \ldots, y_m$ are the dual variables, and if x_j is a basic variable, then the jth component $\bar{C}_j$ of $\bar{C}$ is zero. Thus,

$$\sum_{i=1}^{m} a_{ij} y_i = C_j \quad (5.40)$$

The system in Eqn. (5.40) has a unique solution if rank of $A = m$, since the basis matrix S resulting from m basic variables in A is a non-singular column vector. Let C_s be the basic variables' coefficient in the objective function assuming the rank of $A = m$. Solving the systems of Eqn. (5.40), gives

$$Y^T = C_s S^{-1} \quad (5.41)$$

Substituting Eqn. (5.41) into the relative cost factor of Eqn. (5.39), we have

$$\bar{C} = C - (C_s S^{-1})^T A$$

$$\bar{C}_j = C_j - (C_s S^{-1})^T A_j \quad \text{for the } j^{th} \text{ column of A,}$$

$$= C_j - C_s Y_j \quad \text{since } Y_j = S^{-1} A_j \quad \text{for } j = 1, 2, \ldots, n$$

$$= C_j - Z_j \quad \text{since } Z_j = C_s Y_j \quad \text{for } j = 1, 2, \ldots, n$$

The value of the objective function in terms of basic and non-basic variables can be obtained from Eqns. (5.39) and (5.41) as:

$$Z = C_s S^{-1} B + \bar{C} X \tag{5.42}$$

The relative cost factor of the non-basic variable x_j is the rate of change of the function Z given by

$$\frac{\partial Z}{\partial x_j} = \bar{C}_j = C_j - Z_j \tag{5.43}$$

Now, lets consider the jth dual constraint for a maximization type of primal problem in the context of the above discussion as follows:

$$a_{1j} y_1 + a_{2j} y_2 + \ldots + a_{mj} y_m \geq C_j \tag{5.44}$$

where a_{ij} is the per unit measurement of the jth primal variables x_j from the ith right-hand side vector b_i, and C_j is the objective function coefficient of x_j in terms of basic and non-basic variables given in Eqn. (5.42). The optimality criteria suggests that x_j becomes the promising variable provided $Z_j - C_j \leq 0$. Since the primal is maximizing, C_j may be considered as benefit or profit while $-Z_j$ may be regarded as the cost or loss. Economically, Z_j may be taken as an imputed price per unit of x_j and the dual variables $y_i's$ are the worth per unit measurement of a_{ij}. If the per unit profit C_j exceeds its per unit imputed price Z_j, then the non-basic variable x_j remains the promising variable for the optimal solution, i.e., $C_j \geq Z_j$ or $C_j - Z_j \leq 0$. The present value of the objective function does not change by the corresponding variables when $Z_j = C_j$. However, when $Z_j - C_j \geq 0$, the variable x_j remains non-basic at zero level. It can be easily concluded that for a basic variable $Z_j = C_j$ means that the optimal solution cannot be improved further by the corresponding variable.

5.8.1 Conversion of Primal to Dual

The term "Dual" in general implies two or double. In LP, duality implies that corresponding to every LPP there exists another LPP called its dual. The following pair of LPPs are the standard primal-dual pair in canonical (or symmetrical) form:

min $Z = c' x$	max $Z' = b' w$
subject to $Ax \geq b$	subject to $A' w \leq c$
and $x \geq 0$	and $w \geq 0$

To write the dual of any given LPP following steps can be followed:

(i) Convert the LPP in the standard canonical form leaving equations and unrestricted variables as they are.

(ii) Write down the numerical values of the parameters $A, \underline{b}$ & $\underline{c}$. Also, obtain A' and $\underline{b}'$.

(iii) Define the dual vector $\underline{w}$. The number of components in $\underline{w}$ is equal to the primal constraints.

(iv) Write down the dual LPP using the standard primal-dual pair with the following:

  (a) If any primal constraint is an equation the corresponding dual variable will become unrestricted in sign.

  (b) If any primal variable is unrestricted in a sign, the corresponding dual constraint will become an equation.

Example 66 *Write the dual of the given LPP:*

$$\begin{aligned} \max \ Z &= 6x_1 + 8x_2 \\ \text{subject to } x_1 - 5x_2 &\leq 24 \\ 3x_1 + 7x_2 &\leq 15 \\ \text{and } x_1, x_2 &\geq 0 \end{aligned} \tag{5.45}$$

Following the steps of dual, the dual of the LPP (5.45) is represented by (5.46):

$$\begin{aligned} \min \ Z &= 24w_1 + 15w_2 \\ \text{subject to } w_1 + 3w_2 &\geq 6 \\ -5w_1 + 7w_2 &\geq 8 \\ \text{and } w_1, w_2 &\geq 0 \end{aligned} \tag{5.46}$$

Example 67 *Write the dual of the given LPP:*

$$\begin{aligned} \min \ Z &= 5x_1 - 2x_2 + 6x_3 \\ \text{subject to } 2x_1 + 4x_2 + 3x_3 &\geq 8 \\ 6x_1 + x_2 + 3x_3 &\geq 4 \\ 2x_1 - 7x_2 - 2x_3 &\leq 10 \\ 3x_1 - x2 + 4x_3 &\geq 3 \\ 2x_1 + 5x_2 - x_3 &\geq 2 \\ \text{and } x_1, x_2, x_3 &\geq 0 \end{aligned} \tag{5.47}$$

Since the problem is of minimization, therefore, all the constraints (inequalities) have changed to $\geq$ type by multiplying both sides by (-1). Hence the third constraint will now become $-2x_1 + 7x_2 + 2x_3 \geq -10$. Now, following the steps of dual, the dual of the LPP (5.47) is represented by (5.48):

$$\max \ Z = 8w_1 + 4w_2 - 10w_3 + 3w_4 + 2w_5$$
$$\text{subject to } 2w_1 + 6w_2 - 2w_3 + 3w_4 + 2w_5 \leq 5$$
$$4w_1 + w_2 + 7w_3 - w_4 + 5w_5 \leq -2 \qquad (5.48)$$
$$3w_1 + 3w_2 + 2w_3 + 4w_4 - w_5 \leq 6$$
$$\text{and } w_1, w_2, w_3, w_4, w_5 \geq 0$$

Example 68 *Obtain the dual LPP of the following primal LPP:*

$$\min \ Z = x_1 + 5x_2$$
$$\text{subject to } 3x_1 + 2x_2 \leq 150$$
$$x_1 - x_2 = 40 \qquad (5.49)$$
$$x_1 \geq 15$$
$$\text{and } x_1, x_2 \geq 0$$

Since the problem is of minimization, therefore, all the constraints (inequalities) have changed to $\geq$ type by multiplying both sides by (-1). Hence the first constraint will now become $-3x_1 - 2x_2 \geq -150$. Now, following the steps of dual, the dual of the LPP (5.49) is represented by (5.50):

$$\max \ Z = -150w_1 + 40w_2 + 15w_3$$
$$\text{subject to } -3w_1 + w_2 + w_3 \leq 1$$
$$-2w_1 - w_2 \leq 5 \qquad (5.50)$$
$$\text{and } w_1, w_3 \geq 0, \quad w_2 \ \text{unrestricted}$$

The second constraint of the problem (5.49) is an equality, therefore, the second variable in the dual problem (5.50) is unrestricted.

Example 69 *Obtain the dual LPP of the following primal LPP:*

$$\min \ Z = x_1 - 3x_2$$
$$\text{subject to } 3x_1 - 2x_2 \leq 5$$
$$2x_1 - x_2 \geq 10 \qquad (5.51)$$
$$-4x_1 + 3x_2 \geq 10$$
$$\text{and } x_1 \geq 0 \quad x_2 \ \text{unrestricted}$$

Since the problem is of minimization, therefore, all the constraints (inequalities) have changed to $\geq$ type by multiplying both sides by (-1). Hence the first constraint will now become $-3x_1 + 2x_2 \geq -5$. Now, following the steps of dual, the dual of the LPP (5.51) is represented by (5.52):

$$\max \ Z = -5w_1 + 10w_2 + 10w_3$$
$$\text{subject to } -3w_1 + 2w_2 - 4w_3 \leq 1$$
$$2w_1 - w_2 + 3w_3 = -3 \quad (5.52)$$
$$\text{and } w_1, w_2, w_3 \geq 0$$

The second variable of the problem (5.51) is unrestricted, therefore, the second constraint in the dual problem (5.52) is an equality.

5.8.2 Importance of Duality Concepts

Suppose if we have an LPP with a large number of constraints and relatively few variables, then its dual will have a large number of variables and fewer constraints. In such cases it is feasible to solve the dual rather than solving the primal problem because an additional constraint requires more computational effort than an additional variable.

Based on the above discussion it can be concluded that in such cases dual LPP is easy for solving while primal LPP is quite complex.

Theorem 5.2
Complementary Slackness Theorem: If $\underline{x}^*$ and $\underline{w}^*$ are feasible to primal and dual LPP, in canonical form, then they are optimal to primal and dual respectively if and only if

$$(c_j - \underline{w}^* \underline{a}_j) x_j^* = 0; \ j = 1, 2, \ldots, n$$
$$\text{and } \underline{w}^* (\underline{a}^i \underline{x}^* - b_i) = 0; \ i = 1, 2, \ldots, n$$

Conversely, optimal vectors must satisfy complementary slackness conditions.

Example 70 *Consider the following LPP*

$$\min \ Z_P = 2x_1 + 3x_2 + 4x_3$$
$$\text{subject to } x_1 + 2x_2 + x_3 \geq 3$$
$$2x_1 - x_2 + 3x_3 \geq 4 \quad (5.53)$$
$$\text{and } x_1, x_2, x_3 \geq 0$$

Define the dual vector $\underline{w} = (w_1, w_2)$. Then, the dual LPP is

$$\max \ Z_D = 3w_1 + 4w_2$$
$$\text{subject to } w_1 + 2w_2 \leq 2$$
$$2w_1 - w_2 \leq 3 \quad (5.54)$$
$$w_1 + 3w_2 \leq 4$$
$$\text{and } w_1, w_2 \geq 0$$

Solving the dual LPP we get the optimal solution as $w_1^* = 8/5, w_2^* = 1/5$ and $Z_D^* = 28/5$. Using the complementary slackness property, since both the dual variables w_1 & w_2 are greater than zero (≥ 0) at the optimal point both the primal constraints will be active at the primal optimal point $\underline{x}^* = (x_1^*, x_2^*, x_3^*)$.

Furthermore, since the third dual constraint is inactive at the dual optimal point, the third primal variable $x_3^* = 0$ at the primal optimal point.

$\Rightarrow$ The primal constraints will reduce to

$$x_1^* + 2x_2^* = 3$$
$$2x_1^* - x_2^* = 4$$

This gives $x_1^* = 11/5$ and $x_2^* = 2/5$.

Thus the primal optimal solution is $x_1^* = 11/5, x_2^* = 2/5, x_3^* = 0$ and the optimal objective function value is $Z_P^* = 28/5 = Z_D^*$ same as the dual objective function value.

Note:

1. Primal optimality implies dual feasibility. As primal and dual are interchangeable, therefore, we can also say dual feasibility implies primal optimality.

2. $\underline{w}^* = \underline{c}_B B^{-1}$ is an optimal solution to the dual because $\underline{b}\underline{w}^* = \underline{c}_B B^{-1} \underline{b} = \underline{c}_B \underline{x}_B^*$, where $\underline{x}^*$ is the optimal solution to the primal with the optimal value of the objective function $Z_P^* = \underline{c}_B \underline{x}_B^* = \underline{b}\underline{w}^* = Z_D^*$ at $\underline{w}^*$.

Theorem 5.3
Duality Theorem: *If either the primal problem or the dual problem has a finite optimal solution, then the other problem has a finite optimal solution as well with the same objective function values, that is, the minimum of $c'\underline{x}$ is equal to the maximum of $b\underline{w}$.*

Theorem 5.4
Weak Duality Theorem:
Consider the primal LPP as:

$$\max \ Z_P = c\underline{x}$$
$$\text{subject to } A\underline{x} \leq \underline{b} \quad (5.55)$$
$$\text{and } \underline{x} \geq 0$$

with $F_P = \{\underline{x} | A\underline{x} \leq \underline{b}, \underline{x} \geq 0\}$ be the set of all feasible solution for the primal LPP.

Then, the corresponding dual LPP is:

$$\min \ Z_D = b^T \underline{w}$$
$$\text{subject to } A^T \underline{w} \geq \underline{c}^T \tag{5.56}$$
$$\text{and } \underline{w} \geq 0$$

with $F_D = \{\underline{w} | A^T \underline{w} \geq \underline{c}^T, \underline{w} \geq 0\}$ being the set of all feasible solutions for the dual problem.

If primal and dual problem are both feasible with F_P and F_D as their respective feasible regions then, $Z_P \leq Z_D$ for every $\underline{x} \in F_P$ and $\underline{w} \in F_D$.

Proof 5.2 Let $\underline{x}_0 \in F_P$ and $\underline{w}_0 \in F_D$. Accordingly, we have

$$A\underline{x}_0 \leq \underline{b} \tag{5.57}$$

$$\underline{x}_0 \geq 0 \tag{5.58}$$

and

$$A^T \underline{w}_0 \leq \underline{c}^T \ \text{ or } \ \underline{w}_0^T A \geq \underline{c} \tag{5.59}$$

$$\underline{w}_0^T \geq 0 \tag{5.60}$$

In view of Eqns. (5.57)-(5.60), we have

$$\underline{w}_0^T A \underline{x}_0 \leq \underline{w}_0^T \underline{b} \tag{5.61}$$

and from Eqns. (5.58)-(5.59)

$$\underline{w}_0^T A \underline{x}_0 \leq \underline{c} \underline{x}_0 \tag{5.62}$$

From Eqns. (5.61) and (5.62)

$$\underline{c}\underline{x}_0 \leq \underline{w}_0^T A \underline{x}_0 \leq \underline{w}_0^T \underline{b}$$
$$\text{or } \underline{c}\underline{x}_0 \leq \underline{w}_0^T A \underline{x}_0 \leq b^T \underline{w}_0$$
$$\Rightarrow Z_P \leq Z_D$$

Example 71 *Consider a primal LPP*

$$\max \ Z_P = 2x_1 + 3x_2$$
$$\text{subject to } 5x_1 + x_2 \leq 7$$
$$3x_1 + x_2 \leq 11 \tag{5.63}$$
$$\text{and } x_1, x_2 \geq 0$$

and the dual LPP is:

$$\min \ Z_D = 7w_1 + 11w_2$$
$$\text{subject to } 5w_1 + 3w_2 \geq 2$$
$$w_1 + w_2 \geq 3 \tag{5.64}$$
$$\text{and } w_1, w_2 \geq 0$$

The feasible solution for the primal LPP is $x_1 = 0$ and $x_2 = 6$ which gives $Z_P = 17$ and $w_1 = 3$ and $w_2 = 0$ are feasible solutions for the dual LPP with value $Z_D = 21$. It can be seen that $Z_P \leq Z_D$, which verifies the statement of weak duality theorem.

5.8.3 Properties of Primal-dual LPPs

1. The dual of the dual is primal.
2. If $\underline{x}_0$ and $\underline{w}_0$ be feasible to primal and dual respectively, then $\underline{c x}_0 \geq \underline{w}_0 \underline{b}$.
3. If $\underline{x}_*$ and $\underline{w}_*$ be feasible to the primal and dual respectively and $\underline{c x}_* = \underline{w}_* \underline{b}$, then $\underline{x}_*$ and $\underline{w}_*$ are optimal to primal and dual respectively.
4. If one problem has an unbounded solution, then the other problem is infeasible (no solution).
5. If one problem has no solution (infeasible) then the other problem is either unbounded or infeasible.
6. If either the primal or the dual problem has a finite optimum solution then the other problem has a finite optimum solution, and the optimum values of the two objective functions are equal.

5.8.4 Economic Interpretation of Duality

Consider the problem of production planning. The production manager attempts to determine quantities for each product to be manufactured to optimize the use of available resources to maximize profit. But through a dual LP problem approach, he may develop a production plan that optimizes resource utilization so that the marginal opportunity cost of each unit of a resource is equal to its marginal return (also known as Shadow price).

The shadow price indicates an additional price to be paid to obtain one additional unit of the resources to maximize profit under the resource constraints.

$$Shadow\ Price = \frac{Change\ in\ optimal\ objective\ function\ value}{Unit\ change\ in\ the\ availability\ of\ resource}$$

Example 72 *An XYZ company manufactures two types of products A & B, using two different machines. The time required to manufacture each product by each machine is given in Table* (5.9). *Both products A & B contribute a profit of $6 and $5 respectively. How many units of each product will XYZ produce to attain a maximum profit in each production run? What is the economic interpretation of the optimal values for the primal and dual variables?*

Table 5.9: Manufacturing Time.

Machines	Machine time in hours for producing a unit of products A & B		Maximum available time in hours/week
	A	B	
I	1	2	10
II	3	1	20

Let's define x_1 and x_2 as the number of units for products A & B (primal variables) and Z the total profit. The objective of XYZ is to maximize the total profit Z. Thus,

$$\max \ Z = 6x_1 + 5x_2$$

The machine I takes 1 hour to manufacture product A and 2 hours for product B, while Machine II utilizes 3 hours in manufacturing product A and 1 hour for B. The weekly time restriction of machines I and II are 10 and 20 hours, respectively. These restrictions can be written mathematically as:

$$x_1 + 2x_2 \leq 10;$$

$$3x_1 + x_2 \leq 20;$$

The silence constraint is that neither product A nor B can be negative. That is, no negative production, which implies that

$$x_1, x_2 \geq 0;$$

The above formulation is the primal problem. Solving the primal, the optimal solution is $x_1 = 6$ units, $x_2 = 2$ units, and $Z = \$46$.

Similarly, let's define the dual variables as y_1, y_2, and ω as the total time used in production. Thus, a dual to the given primal LP can be formulated as:

$$\begin{aligned} \min \ \omega &= 10y_1 + 20y_2 \\ \text{subject to } y_1 + 3y_2 &\geq 6 \\ 2y_1 + y_2 &\leq 5 \\ \text{and } y_1, y_2 &\geq 0 \end{aligned} \quad (5.65)$$

The optimal solution of the dual is $y_1 = 1.8$ hours, $y_2 = 1.4$ hours, and $\omega = \$46$; The economic interpretation of the dual problem is as follows:

- It can be seen from the solution that the $\min \omega = \max Z$.

- Since the coefficients of the primal objective function Z are the associated profits in $, the right-hand side of the dual problem must also be in $.

- The RHS of the primal problem constraints are in hours. So, they become constant coefficients in the dual objective function.

- The dual variables are the shadow price or the incurred costs (implicit price) in $ for an hour utilization for both machines.

The strong duality theorem suggests that the set of activities and price must have an equilibrium, i.e., the maximum profit from production equals the minimum rental cost. Now, the optimal price can be interpreted as follows:

- If machine II operates only for 20 hours and the primal optimal basis does not change, then the optimal value of Z increases by $1.8 if the time available for the machine I increases by $y_1 = 1.8$ hours (i.e., from 10 hours to 11.8 hours).

- Similarly, the optimal value of Z will increase by $1.4 if the available time on machine II increases by $y_2 = 1.4$ hours, provided that the operation time for the machine I is only 10 hours and the optimal primal basis does not change.

5.8.5 Dual Simplex

In the usual simplex method, we start with a feasible but non-optimal solution and maintaining its feasibility iteration by iteration and improving it towards optimality. In the dual simplex method, we start with an optimal but infeasible solution and maintain its optimality iteration by iteration that improves it towards feasibility.

The various steps of the dual simplex method for a minimization LPP are given below:

Step 1 Find a basis B such that

$$z_j - c_j = \underline{c}'_B B^{-1} \underline{a}_j - c_j \leq 0 \ \forall \ j \quad (5.66)$$

Prepare the usual simplex tableau.

Step 2 If the RHS vector $\overline{b} = B^{-1}\underline{b} \geq 0$, STOP; the current solution is optimum. Otherwise select the lth row as the pivotal row where

$$\overline{b}_l = \min_i \{\overline{b}_i\}$$

$\overline{b}_i$ is the ith element of the RHS vector $\overline{b}$.

Step 3 If $y_{lj} \geq 0$ for all j, STOP, the given LPP has no solution, where y_{lj} is the jth element of the pivotal row. Otherwise select the kth column as the pivotal column where

$$\frac{z_k - c_k}{y_{lk}} = \min_{y_{lj} < 0} \left\{ \frac{z_j - c_j}{y_{lj}} \right\} \quad (5.67)$$

y_{lk} will be the pivotal element. RHS of Eqn. (5.67) gives the minimum positive ratio.

Step 4 Construct the next simplex tableau as usual. Repeat step (2) to (4).

Note: For a maximization LPP expressions (5.66) and (5.67) at steps 1 and 3 will change to:

$$z_j - c_j = \underline{c}'_B B^{-1} \underline{a}_j - c_j \geq 0 \;\; \forall \;\; j \tag{5.68}$$

$$\frac{z_k - c_k}{y_{lk}} = \max_{y_{lj} < 0} \left\{ \frac{z_j - c_j}{y_{lj}} \right\} \tag{5.69}$$

respectively. RHS of Eqn. (5.69) gives the maximum negative ratio.

Example 73 *Solve the following LPP by using the dual simplex method.*

$$\begin{aligned} \min \;\; & Z = 5x_1 + 6x_2 \\ \text{subject to} \;\; & x_1 + x_2 \geq 2 \\ & 4x_1 + x_2 \geq 4 \\ \text{and} \;\; & x_1, x_2 \geq 0 \end{aligned} \tag{5.70}$$

Solution Step 1: Standard Form- To convert the inequality ($\leq$) constraints to equality, slack variables s_1 and s_2 are added in each constraint. Then, the standard LPP is:

$$\begin{aligned} \max \;\; & Z = -5x_1 - 6x_2 + 0s_1 + 0s_2 \\ \text{subject to} \;\; & -x_1 - x_2 + s_1 = -2 \\ & -4x_1 - x_2 + s_2 = -4 \\ \text{and} \;\; & x_1, x_2, s_1, s_2 \geq 0 \end{aligned} \tag{5.71}$$

Step 2: The initial basic feasible solution is corresponding to s_1 and s_2, therefore

$$B = \begin{bmatrix} 1 & 0 \\ 0 & 1 \end{bmatrix}$$

Step 3:

$$B^{-1} = \begin{bmatrix} 1 & 0 \\ 0 & 1 \end{bmatrix}$$

$$B^{-1} \underline{b} = \begin{bmatrix} 1 & 0 \\ 0 & 1 \end{bmatrix} \begin{bmatrix} 4 \\ 2 \end{bmatrix} = \begin{bmatrix} 4 \\ 2 \end{bmatrix}$$

Step 4: The initial simplex tableau is:

All $z_j - c_j \geq 0$ but all $x_{Bi} \not\geq 0$, therefore the current basic feasible solution is not optimal and we have to update the solution. Now x_1 corresponding to the most negative value of $z_j - c_j$ will enter the basis. To identify the leaving variable the minimum ratio between x_B and the key column has been calculated and it is found that s_2 leaves the basis as the minimum ratio $5/-4 = -1.22$ is corresponding to s_2.

Linear Optimization Problems ■ 113

Table 5.10: Tableau '0'.

		c_j	-5	-6	0	0
c_B	Basis	x_B	x_1	x_2	s_1	s_2
0	s_1	-2	-1	-1	1	0
←0	s_2	-4	-4	-1	0	1
		z_j	0	0	0	0
		$z_j - c_j$	5↑	6	0	0

To update the solution convert the key element, i.e., -4 into unity by dividing the key row by the key element and the rest of the other elements in the key column to zero by using the formula given below:

$$New\ element = \left\{old\ element - \frac{Product\ of\ elements\ in\ key\ row\ and\ column}{key\ element}\right\}$$

All $z_j - c_j \geq 0$ but all $x_{Bi} \not\geq 0$, therefore the current basic feasible solution is not optimal and we have to again update the solution following the same procedure. Now the entering variable is s_2 and the leaving variable is s_1.

Table 5.11: Iteration '1'.

		c_j	-5	-6	0	0
c_B	Basis	x_B	x_1	x_2	s_1	s_2
←0	s_1	-1	0	-3/4	1	-1/4
-5	x_1	1	1	1/4	0	-1/4
		z_j	-5	-5/4	0	5/4
		$z_j - c_j$	0	19/4	0	5/4↑

Table 5.12: Iteration '2'.

		c_j	-5	-6	0	0
c_B	Basis	x_B	x_1	x_2	s_1	s_2
0	s_2	4	0	3	-4	1
-5	x_1	2	1	1	-1	0
		z_j	-5	-5	5	0
		$z_j - c_j$	0	1	5	0

Since all $z_j - c_j \geq 0$ and also all $x_{Bi} \geq 0$, the current solution is optimum. The optimal solution is $x_1 = 2$ and $x_2 = 0$ with Max $Z = -10$, or Min $Z = 10$. ■

Now, if we solve the same problem in LINGO-18 then in one click we get the same result. The LINGO-18 codes for the problem (5.70) are:

LINGO-18 codes for the problem (5.70)

MODEL :
SETS :
product/1..2/ : c,x;
resource/1..2/ : mat;
material(resource, product) : a;
endsets
!minimization objective function;
[objective]min = @sum(product(j) : c(j) * x(j));
!constraints;
@for(capacity(i) : [material_constraints]
@sum(product(j) : a(i, j) * x(j)) >= mat(i));
Data :
c = 5, 6;
mat = 2, 4;
a = 1 1 4 1;
enddata
end

The global optimal solution of LPP (5.70) derived through LINGO-18 software is $x_1 = 2$, $x_2 = 0$ with minimum Z = 10.

Chapter 6
Non-Linear Optimization Problems

6.1 Introduction

An optimization problem in which all the involved functions are not linear is called a non-linear optimization problem or non-linear programming problem (NLPP). The mathematical model of an NLPP may be given as:

$$\left.\begin{array}{l} \max(\min) \ = f(\underline{x}) \\ \text{subject to } g_i(\underline{x})(\leq,=,\geq)b_i; \ i=1,2,\ldots,m \\ \text{and } \underline{x} \geq 0 \end{array}\right\} \quad (6.1)$$

where $\underline{x}$ is an n-component vector of decision variables $x_1, x_2, \ldots, x_n$ and at least one of the $m+1$ functions $f(\underline{x}), g_1(\underline{x}), g_2(\underline{x}), \ldots, g_m(\underline{x})$ is not linear.

For developing the theory, the following form of an NLPP may be considered as standard form:

$$\left.\begin{array}{l} \max \ = f(\underline{x}) \\ \text{subject to } g_i(\underline{x}) \geq b_i; \ i=1,2,\ldots,m \\ \text{and } \underline{x} \geq 0 \end{array}\right\} \quad (6.2)$$

6.1.1 Basic Definitions

Definition 6.1 **Feasible Solution:** An n-component vector $\underline{x}$ is called a feasible solution to the NLPP (6.2) if it satisfies the constraints $g_i(\underline{x}) \geq b_i; \ i=1,2,\ldots,m$ and

$\underline{x} \geq 0$. The set F of all feasible solutions to NLPP (6.2) is defined as:

$$F = \{\underline{x} | g_i(\underline{x}) \geq 0;\ i = 1, 2, \ldots, m \text{ and } \underline{x} \geq 0\}$$

Definition 6.2 **Optimal Solution:** An $\underline{x}^* \in F$ is called an optimal solution to NLPP (6.2) if

$$f(\underline{x}^*) \geq f(\underline{x})\ \forall\ \underline{x} \in F$$

Definition 6.3 **Feasible Direction:** A direction $\underline{d}$ at a feasible point $\underline{x}$ is called a feasible direction if $(\underline{x} + \gamma \underline{d}) \in F$ for a sufficiently small γ. The set of all feasible directions $\underline{d}$ at a point $\underline{x} \in F$ is defined as:

$$D(x) = \{\underline{d} | (\underline{x} + \gamma \underline{d}) \in F\ \forall\ 0 \leq \gamma \leq \sigma;\ \sigma > 0\}$$

Definition 6.4 **Directional Derivative:** The directional derivative of $f(\underline{x})$ along the direction $\underline{d}$ is given by

$$\lim_{\gamma \to 0} \frac{f(\underline{x} + \gamma \underline{d}) - f(\underline{x})}{\gamma}$$

Using Taylor's theorem it can be seen that

$$\lim_{\gamma \to 0} \frac{f(\underline{x} + \gamma \underline{d}) - f(\underline{x})}{\gamma} = [\nabla f(\underline{x})]' \underline{d}$$

Definition 6.5 **Closure of a set:** The closure of $S \subset E_n$ denoted by $\overline{S}$, is the set of all points that are arbitrarily close to S, that is if $\underline{x} \in \overline{S}$ then for each $\varepsilon > 0$, however small, $S \cap N_\varepsilon(\underline{x}) \neq \phi$, where $N_\varepsilon(\underline{x})$ denote the ε-neighbourhood of $\underline{x}$.

Definition 6.6 **A closed set:** A set S is called closed if $S = \overline{S}$.

Definition 6.7 **Active and inactive constraints at any feasible point:** Any feasible point $\underline{x}$ for the kth constraint is said to be active if $g_k(\underline{x}) = 0$ and inactive if $g_k(\underline{x}) > 0$. Thus at any feasible point the constraints may be divided into two disjoint sets of active and inactive constraints.

Without loss of generality, we can assume that at any feasible $\underline{x}$ the first l constraints are active and remaining $m - l$ constraints are inactive, that at any $\underline{x} \in F$

$$g_i(\underline{x}) = 0;\ i = 1, 2, \ldots, l$$

$$\text{and } g_i(\underline{x}) > 0;\ i = l+1, l+2, \ldots, m$$

Also, for any feasible point $\underline{x}$, some components of $\underline{x}$ may be zero. Thus without loss of generality, we may assume that

$$x_j = 0;\ j = 1, 2, \ldots, k$$

and $x_j > 0$; $j = k+1, k+2, \ldots, n$

Now, define the set of direction $\underline{d}$ as:

$$D(\underline{x}) = \{\underline{d} \mid \nabla g_i(\underline{x})' \underline{d} \geq 0;\ i = 1, 2, \ldots, l\ \&\ d_j \geq 0;\ j = 1, 2, \ldots, k\}$$

It can be verified that the set $D(\underline{x})$ is a closed set.

6.1.2 Some Properties

1. Let f be differentiable at $\underline{x}$, then the $\nabla f(\underline{x})$, if it is not zero, points in a direction such that a small movement along that direction will increase f.
2. If $\underline{x}^*$ is optimal for NLPP (6.2) then $\nabla f'(\underline{x}^*)\underline{d} \leq 0$ for all $\underline{d} \in D(\underline{x}^*)$.
3. If $\underline{x}^*$ is optimal for NLPP (6.2) then $\nabla f'(\underline{x}^*)\overline{\underline{d}} \leq 0$ for all $\overline{\underline{d}} \in \overline{D}(\underline{x}^*)$.
4. $\overline{D}(\underline{x})$ is a proper subset of $D(\underline{x})$ that is for a feasible direction at $\underline{d}$ at a feasible point $\underline{x}$ we have

$$\nabla' g_i(\underline{x})\underline{d} \geq 0;\ i = 1, 2, \ldots, l \text{ and } d_j > 0;\ j = 1, 2, \ldots, k$$

<u>Constraint Qualification</u>: If $\underline{x}^*$ is the optimal solution, then the constraint qualification imposes the following restriction on the constraint set

$$D(\underline{x}^*) \subset \overline{D}(\underline{x}^*)$$

Using property 4 we have

$$D(\underline{x}^*) = \overline{D}(\underline{x}^*)$$

6.2 Lagrange Multiplier

When an NLPP is composed of some differentiable functions and constraints are of equality, in such a situation, an optimal solution can be achieved through Lagrange Multipliers. It helps in calculating the sensitivity of the best value of the objective function to the change in the given constraints b_i in the problem. Consider an NLPP:

$$\left.\begin{array}{r} \max\ = f(\underline{x}) \\ \text{subject to}\ g_i(\underline{x}) = b_i;\ i = 1, 2, \ldots, m \\ \text{and}\ \underline{x} \geq 0 \end{array}\right\} \quad (6.3)$$

Now, to solve the problem (6.3), the following steps of the Lagrange Multiplier can be followed:

Step 1 Formulate the Lagrange function γ by introducing a new variable λ

$$\gamma(x_1,x_2,\ldots,x_n;\lambda_1,\lambda_2,\ldots,\lambda_m) = f(x_1,x_2,\ldots,x_n) + \lambda_i g_i(x_1,x_2,\ldots,x_n)$$

where λ_i; $i = 1,2,\ldots,m$ is the Lagrange Multiplier.

Step 2 Set the gradient of γ equal to the zero vector or find the stationary points of γ

$$\nabla \gamma(x_1,x_2,\ldots,x_n;\lambda) = \underline{0}$$

Step 3 Whichever derived solution $\underline{x}^*$ optimizes the objective function is the optimal solution.

6.3 Kuhn-Tucker Conditions

Kuhn and Tucker derived the following six necessary conditions to be satisfied by the optimal solution to an NLPP. These conditions are also sufficient under certain specified circumstances. When these conditions are necessary as well as sufficient, and we can find an $\underline{x}^*$ satisfying these conditions, then this $\underline{x}^*$ will be the solution to the given NLPP. Thus some of the "well-behaving" NLPP may be solved using the Kuhn-Tucker (K-T) conditions.

Theorem 6.1
Let $\underline{x}^*$ be an optimal solution to the NLPP:

$$\left.\begin{array}{l} \max\ = f(\underline{x}) \\ \text{subject to}\ \ g_i(\underline{x}) \geq b_i;\ \ i = 1,2,\ldots,m \\ \text{and}\ \ \underline{x} \geq 0 \end{array}\right\} \quad (6.4)$$

where the functions are f and g_i; $i = 1,2,\ldots,m$ are differentiable. Assume that the constraint qualification holds.

Then, there exist a vector $\underline{u}^*$ such that

(i) $\nabla_{\underline{x}} \phi(\underline{x}^*,\underline{u}^*) \leq \underline{0}$

(ii) $\underline{x}^{*\prime} \nabla_{\underline{x}} \phi(\underline{x}^*,\underline{u}^*) = 0$

(iii) $\nabla_{\underline{u}} \phi(\underline{x}^*,\underline{u}^*) \geq \underline{0}$

(iv) $\underline{u}^{*\prime} \nabla_{\underline{u}} \phi(\underline{x}^*,\underline{u}^*) = 0$

(v) $\underline{x}^* \geq \underline{0}$ *and,*

(vi) $\underline{u}^* \geq \underline{0}$

where $\phi(\underline{x},\underline{u}) = f(\underline{x}) + \sum_{i=1}^{m} u_i g_i(\underline{x})$.

Theorem 6.2
Sufficiency of K-T conditions: If f is pseudo concave and g_i; $i = 1, 2, \ldots, m$ are quasi concave and $\underline{x}^*$ satisfies the K-T necessary conditions, the $\underline{x}^*$ is optimal for NLPP (6.2).

6.4 Solution of Non-Linear Optimization Problems using LINGO-18

Example 74 *Consider a stratified non-linear optimization problem:*

$$\begin{aligned}
\min V &= \frac{652.678}{n_1} + \frac{146.4730}{n_2} + \frac{172.3015}{n_3} + \frac{2677.5213}{n_4} \quad \text{(Sampling Variance)} \\
\text{subject to } & 2n_1 + 5n_2 + 4n_3 + 6n_4 \leq 5500 \quad \text{(Cost Constraint)} \\
& 2 \leq n_1 \leq 1319 \quad \text{(Sample Size limitation Constraints)} \\
& 2 \leq n_2 \leq 720 \\
& 2 \leq n_3 \leq 1153 \\
& 2 \leq n_4 \leq 879 \\
& \text{and } n_h \text{ are integers, } h = 1, 2, 3, 4
\end{aligned} \quad (6.5)$$

To derive the optimal solution of the problem (6.5), the LINGO-18 codes are as follows:

```
MODEL:
MIN= 652.678 / N1 + 146.473 / N2 + 172.3015 / N3 + 2677.5213 / N4;
2 * N1 + 5 * N2 + 4 * N3 + 6 * N4 <= 5500;
@GIN( N1); @BND( 2, N1, 1319); @GIN( N2); @BND( 2, N2, 720);
@GIN( N3); @BND( 2, N3, 1153); @GIN( N4); @BND( 2, N4, 879);
END
```

The optimal solution of the problem (6.5) is $n_1 = 460$, $n_2 = 138$, $n_3 = 167$, $n_4 = 537$ with minimum variance = 8.498082.

Example 75 *Consider a reliability non-linear optimization problem:*

$$\begin{aligned}
\max R_s &= (1 - (1 - 0.99)^{y_{11}} (1 - 0.95)^{y_{12}} (1 - 0.92)^{y_{13}}) \\
& (1 - (1 - 0.98)^{y_{21}} (1 - 0.8)^{y_{22}} (1 - 0.90)^{y_{23}}) \\
& (1 - (1 - 0.98)^{y_{31}} (1 - 0.92)^{y_{32}}) \quad \text{(System Reliability)} \\
\text{subject to } & 3y_{11} + 12y_{12} + 8y_{13} + 7y_{21} + 4y_{22} + 3y_{23} + 11y_{31} + 6y_{32} \leq 40 \quad \text{(Cost Constraint)} \\
& 3y_{11} + 4y_{12} + 6y_{13} + 2y_{21} + 3y_{22} + 7y_{23} + 5y_{31} + 5y_{32} \leq 16 \quad \text{(Weight Constraint)} \\
& y_{11} + y_{12} + y_{13} + y_{21} + y_{22} + y_{23} + y_{31} + y_{32} \geq 1 \\
& \text{and } y_{ij} \geq 0
\end{aligned} \quad (6.6)$$

To derive the optimal solution of the problem (6.6), the LINGO-18 codes are as follows:

```
MODEL:
MAX= ( 1 - ( 1 - 0.99) ^ Y11 * ( 1 - 0.95) ^ Y12 * ( 1 - 0.92) ^ Y13)
* ( 1 - ( 1 - 0.98) ^ Y21 * ( 1 - 0.8) ^ Y22 * ( 1 - 0.9) ^ Y23)
* ( 1 - ( 1 - 0.98) ^ Y31 * ( 1 - 0.92) ^ Y32);
3 * Y11 + 12 * Y12 + 8 * Y13 + 7 * Y21 + 4 * Y22 + 3 * Y23 + 11 * Y31 + 6 * Y32 <= 40;
3 * Y11 + 4 * Y12 + 6 * Y13 + 2 * Y21 + 3 * Y22 + 7 * Y23 + 5 * Y31 + 5 * Y32 <= 16;
Y11 + Y12 + Y13 + Y21 + Y22 + Y23 + Y31 + Y32 >= 1;
END
```

The optimal solution of the problem (6.6) is $y_{11} = 1.49$, $y_{21} = 1.82$, $y_{31} = 1.58$, $y_{12} = y_{13} = y_{22} = y_{23} = y_{32} = 0$ with maximum reliability = 0.99.

Example 76 *Consider a sample surveys non-linear optimization problem in which variance of the population is to be minimized.*

$$\min V = \frac{49}{n_1} + \frac{58.9824}{n_2} + \frac{45.1584}{n_3} + \frac{18.6624}{n_4} + \frac{87.9844}{n_5} \quad \text{(Variance)}$$

subject to $n_1 + n_2 + 1.5n_3 + 1.5n_4 + 2n_5 \leq 1200$ (Cost Constraint)

$2 \leq n_1 \leq 1500$ (Sample Size Bound Constraint)

$2 \leq n_2 \leq 1920$

$2 \leq n_3 \leq 1260$

$2 \leq n_4 \leq 480$

$2 \leq n_5 \leq 840$

and n_j; $j = 1,2,3,4,5$ are integers

(6.7)

To derive the optimal solution of the problem (6.7), the LINGO-18 codes are as follows:

```
MODEL:
MIN= 49 / N1 + 58.9824 / N2 + 45.1584 / N3 + 18.6624 / N4 + 87.9844 / N5;
N1 + N2 + 1.5 * N3 + 1.5 * N4 + 2 * N5 ¡= 1200;
@GIN( N1); @BND( 2, N1, 1500); @GIN( N2); @BND( 2, N2, 1920);
@GIN( N3);
@BND( 2, N3, 1260); @GIN( N4); @BND( 2, N4, 480); @GIN( N5);
@BND( 2, N5, 840);
END
```

The optimal solution of the problem (6.7) is $n_1 = 203, n_2 = 223, n_3 = 158, n_4 = 102, n_5 = 192$ with minimum variance = 1.432904.

Example 77 *Consider a sample surveys non-linear optimization problem in which variance of the population is to be minimized.*

$$\left.\begin{aligned}
\min\ V &= \frac{0.03099}{n_1} + \frac{0.20718}{n_2} \quad \text{(Variance)} \\
\text{subject to }\ & 2n_1 + 20n_2 \leq 4000 \quad \text{(Cost Constraint)} \\
& 1 \leq n_1 \leq 300 \quad \text{(Sample Size Bound Constraint)} \\
& 1 \leq n_2 \leq 7000 \\
\text{and }\ & n_j;\ j = 1, 2 \text{ are integers}
\end{aligned}\right\} \quad (6.8)$$

To derive the optimal solution of the problem (6.8), the LINGO-18 codes are as follows:

```
MODEL:
MIN= 0.03099 / N1 + 0.20718 / N2;
2 * N1 + 20 * N2 <= 4000;
@GIN( N1); @BND( 1, N1, 300); @GIN( N2); @BND( 1, N2, 700);
END
```

The optimal solution of the problem (6.8) is $n_1 = 220, n_2 = 178$ with minimum variance = 0.00131.

Example 78 *Consider a reliability non-linear optimization problem in which the objective is to maximize the reliability of the system under the restrictions of time required to replace the failed components, cost per component, and failed component limitation.*

$$\left.\begin{aligned}
\max\ R &= (1-(1-0.65)^{4+d_1})(1-(1-0.55)^{3+d_2})(1-(1-0.70)^{4+d_3}) \quad \text{(Relaibility)} \\
\text{subject to }\ & 6d_1 + 10d_2 + 7d_3 \leq 75 \quad \text{(Time Constraint)} \\
& 16d_1 + 12d_2 + 13d_3 \leq 150 \quad \text{(Cost Constraint)} \\
& 2 \leq d_1 \leq 3 \quad \text{(Failed Component Constraint)} \\
& 2 \leq d_2 \leq 2 \\
\text{and }\ & 2 \leq d_3 \leq 4
\end{aligned}\right\}$$

(6.9)

To derive the optimal solution of the problem (6.9), the LINGO-18 codes are as follows:

```
MODEL:
MAX= ( 1 - ( 1 - 0.65) ^ ( 4 + D1)) * ( 1 - ( 1 - 0.55) ^ ( 3 + D2)) * ( 1 - ( 1
- 0.7) ^ ( 4 + D3));
6 * D1 + 10 * D2 + 7 * D3 <= 75;
16 * D1 + 12 * D2 + 13 * D3 <= 150;
@BND( 2, D1, 3); @BND( 2, D2, 2); @BND( 2, D3, 4);
END
```

The optimal solution of the problem (6.9) is $d_1 = 3, d_2 = 2, d_3 = 4$ with maximum reliability = 0.98.

Example 79 *Consider a linear plus linear fractional programming problem.*

$$\left. \begin{array}{rl} \min\ Z\ =& (-x_1 - 1) + \dfrac{(-5x_1 + 4x_2)}{(2x_1 + x_2 + 5)} \\ \text{subject to}\ & x_1 - x_2 \geq 2 \\ & 4x_1 + 5x_2 \leq 25 \\ & x_1 + 9x_2 \leq 9 \\ & x_1 \geq 5 \\ \text{and}\ & x_1, x_2 \geq 0 \end{array} \right\} \quad (6.10)$$

To derive the optimal solution of the problem (6.10), the LINGO-18 codes are as follows:

```
MODEL:
MIN= ( - X1 - 1) + ( - 5 * X1 + 4 * X2) / ( 2 * X1 + X2 + 5);
X1 - X2 >= 2;
4 * X1 + 5 * X2 <= 25;
X1 + 9 * X2 >= 9;
X1 >= 5;
END
```

The optimal solution of the problem (6.10) is $x_1 = 5.81$, $x_2 = 0.35$ with minimum objective value = -8.434.

6.5 Quadratic Programming

A Quadratic Programming Problem (QPP) is an NLPP whose objective function is a sum of a linear and a quadratic form, and the constraints are linear. The standard mathematical model of a QPP may be given as:

$$\left.\begin{array}{l}\max\ Q(\underline{x})\ =\underline{c}'\underline{x}+\underline{x}'D\underline{x}\\ \text{subject to}\ A\underline{x}\le \underline{b}\\ \text{and}\ \underline{x}\ge 0\end{array}\right\} \quad (6.11)$$

where D is a $(n \times n)$ symmetric matrix, and all other symbols are as used in the LPP.

Since QPP is a widely used NLPP, therefore various methods are developed for the solution of the QPP under the assumption that the quadratic form $\underline{x}'D\underline{x}$ is concave (for minimization $\underline{x}'D\underline{x}$ is assumed to be convex). Some of the most widely used and accepted methods are Dantzig's, Lemke's, Wolfe's, and Beale's. Out of these methods Dantzig's, Lemke's, and Wolfe's are directly based on the solutions of K-T conditions whereas Beale's method is based on the modified simplex method.

In the next sections, the most popular two methods for solving a QPP are discussed in detail.

6.5.1 Wolf's Method

Consider the QPP as:

$$\left.\begin{array}{l}\max\ Q(\underline{x})\ =\underline{c}'\underline{x}+\underline{x}'D\underline{x}\\ \text{subject to}\ A\underline{x}\le \underline{b}\\ \text{and}\ \underline{x}\ge 0\end{array}\right\} \quad (6.12)$$

Applying Kuhn-Tucker conditions to the QPP (6.12), we get the following set of linear and non-linear equations and the non-negativity restrictions:

$$-2D\underline{x}+A'\underline{\lambda}-\underline{u}=\underline{c} \quad (6.13)$$

$$A\underline{x}+I\underline{s}=\underline{b} \quad (6.14)$$

$$\underline{x},\underline{\lambda},\underline{u},\&\ \underline{s}\ge \underline{0} \quad (6.15)$$

$$\text{and}\ \underline{u}'\underline{x}=\underline{0};\ \underline{\lambda}'\underline{s}=\underline{0} \quad (6.16)$$

Our aim is to find a basic solution to the system of Eqns. (6.13)-(6.14) with (6.15) that satisfies (6.16) also. The $\underline{x}$ part of this solution will solve the QPP.

The required basic solution can be obtained by using 'Phase-I' of the "Two-Phase Simplex Method". The entries in the basis are restricted by the following rule to maintain the conditions imposed in (6.16).

"Rule: If u_j is on the basis at a positive level, x_j cannot become basic at a positive level. Similarly, λ_i and s_i cannot be positive simultaneously."

If Phase-I of the 'Two-Phase Simplex Method' fails to provide a fundamental solution to the system (6.13)-(6.14), then the given QPP will have no solution.

Notes:

- If the QPP is given in minimization form, then convert it into maximization form with $\leq$ type constraints.

- Modify the simplex algorithm to include the complementary slackness conditions.

- The solution is obtained by using Phase I of the simplex method. As the aim is to obtain a feasible solution, there is no need to consider Phase II.

- Phase I ends with the sum of all artificial variables equal to zero, provided that the feasible solution of the problem exists.

Example 80 *Use Wolfe's method to solve the QPP*

$$\left.\begin{array}{l} \max\ Q(\underline{x}) = 4x_1 + 6x_2 - x_1^2 - 3x_2^2 \\ \text{subject to } x_1 + 2x_2 \leq 4 \\ \text{and } x_1, x_2 \geq 0 \end{array}\right\} \quad (6.17)$$

On comparing the QPP (6.17) with the standard QPP (6.12), we get the parameters as:

$$\underline{c}' = (4,6);\ \underline{x} = \begin{pmatrix} x_1 \\ x_2 \end{pmatrix};\ D = \begin{pmatrix} -1 & 0 \\ 0 & -3 \end{pmatrix};\ A = (1,2)\ \text{and } \underline{b} = 4$$

The system of equations

$$-2D\underline{x} + A'\underline{\lambda} - \underline{u} = \underline{c}$$

$$A\underline{x} + I\underline{s} = \underline{b}$$

$$\underline{x}, \underline{\lambda}, \underline{u},\ \&\ \underline{s} \geq \underline{0}$$

$$\text{and } \underline{u}'\underline{x} = \underline{0},\ \underline{\lambda}'\underline{s} = \underline{0}$$

After substituting the values of all parameters, the equations can be expressed as:

$$2x_1 + \lambda_1 - u_1 = 4 \quad (6.18)$$

$$6x_2 + 2\lambda_1 - u_2 = 6 \quad (6.19)$$

$$x_1 + 2x_2 + s_1 = 4 \quad (6.20)$$

$$x_1, x_2, \lambda_1, u_1, u_2, s_1 \geq 0 \quad (6.21)$$

$$u_1 x_1 = u_2 x_2 = 0 = \lambda_1 s_1 \quad (6.22)$$

To apply the Phase-I of the Two-Phase Simplex method, we need two artificial variables $a_1, a_2 \geq 0$, (say). The artificial linear programming problem will be as follows:

$$\left. \begin{array}{l} \min \quad = a_1 + a_2 \\ \text{subject to } 2x_1 + \lambda_1 - u_1 + a_1 = 4 \\ \qquad\qquad 6x_2 + 2\lambda_1 - u_2 + a_2 = 6 \\ \qquad\qquad x_1 + 2x_2 + s_1 = 4 \\ \text{and } x_1, x_2, \lambda_1, u_1, u_2, s_1, a_1, a_2 \geq 0 \end{array} \right\} \qquad (6.23)$$

A starting basic feasible solution will be $a_1 = 4, a_2 = 6, s_1 = 4$ and all other variables equal to zero with identity basis matrix B.

The various simplex tableaus of Phase-I are given below:

Table 6.1: Tableau '1'.

c_B	Basic value	c_j Present Value	0 x_1	0 x_2	0 λ_1	0 u_1	0 u_2	0 s_1	1 a_1	1 a_2
1	a_1	4	2	0	1	-1	0	0	1	0
←1	a_2	6	0	6	2	0	-1	0	0	0
0	s_1	4	1	2	0	0	0	1	0	1
	$z_j - c_j$	2	6↑	3	-1	-1	0	0	0	

From Table (6.1), a_2 will leave the basis as it has minimum ratio $6/6 = 1$ and x_2 will become a basic variable in its place. By the usual transformation formula discussed in Chapter 5, we get the next tableau (Table (6.2)) as:

Table 6.2: Tableau '2'.

c_B	Basic value	c_j Present Value	0 x_1	0 x_2	0 λ_1	0 u_1	0 u_2	0 s_1	1 a_1	1 a_2
←1	a_1	4	2	0	1	-1	0	0	1	0
0	x_2	1	0	1	1/3	0	-1/6	0	0	1/6
0	s_1	2	1	0	-2/3	0	1/3	1	0	-1/3
	$z_j - c_j$	2↑	0	1	-1	0	0	0	0	-1

Now, a_1 will leave the basis and x_1 will enter. The next tableau (Table (6.3)) after the usual transformations is:

Since all $z_j - c_j \leq 0$ and all the artificial variables are non-basic, the solution provided by the Tableau 3 is the required basic solution to the problem. Therefore, the solution is $x_1^* = 2, x_2^* = 1$ and $Q^* = 4 \times 2 + 6 \times 1 - 2^2 - 3 \times 1^2 = 7$.

Table 6.3: Tableau '3'.

c_B	Basic value	c_j Present Value	0 x_1	0 x_2	0 λ_1	0 u_1	0 u_2	0 s_1	1 a_1	1 a_2
0	x_1	2	1	0	1/2	-1/2	0	0	1/2	0
0	x_2	1	0	1	1/3	0	-1/6	0	0	1/6
0	s_1	0	0	0	-7/6	1/2	1/3	1	-1/2	-1/3
	$z_j - c_j$	0	0	0	0	0	0	0	-1	-1

6.5.2 Beale's Method

To achieve the optimum solution of the QPP without using Kuhn-Tucker conditions, a method called Beale's was developed by E.M.L. Beale in 1959. This method developed by him involves the partitioning of the variable into basic and non-basic variables and using traditional calculus results. At every iteration of the simplex method, the objective function is expressed in terms of only the non-basic variables.

Consider the QPP

$$\left.\begin{array}{l} \max\ Z(\underline{x}) = \underline{c}'\underline{x} + \underline{x}'D\underline{x} \\ \text{subject to } A\underline{x} = \underline{b} \\ \text{and } \underline{x} \geq 0 \end{array}\right\} \quad (6.24)$$

It is assumed that the objective function $Z(\underline{x})$ is concave (convex for the minimization problem), and a basic feasible solution to the above QPP exists and is known. It is also assumed that all the basic feasible solutions to the QPP (6.24) are non-degenerate.

6.5.3 Algorithm

The following steps are involved in the algorithm of Beale's method for solving such type of QPP:

Step 1 Introduce slack/surplus variables to convert the linear constraints of the QPP in the equation form.

Step 2 Now divide the total variables into m arbitrarily basic variables and the remaining n variables as non-basic. With this partitioning, the constraint $Ax = b$ can be written as

$$(B \quad R) \begin{bmatrix} x_B \\ x_{NB} \end{bmatrix} = b \text{ or } Bx_b + Rx_{NB} = b$$

where x_B denotes the basic vector and x_{NB} non-basic vector respectively. In addition, A matrix is divided into B & R sub-matrices corresponding to x_B and x_{NB} respectively.

Therefore, the solution can be written as

$$x_B = B^{-1}b - B^{-1}Rx_{NB}$$

Step 3 Using the given and additional constraint equation (if any), write the basic variable x_B in terms of non-basic variable x_{NB} only.

Step 4 Further, using the given and additional constraint (if any), express the objective function $Z(\underline{x})$ also in terms of the non-basic variable x_{NB} only. Thus, the objective function value can be improved by increasing the value of any of the non-basic variables x_{NB}. Now, the constraints of the new problem will be:

$$B^{-1}Rx_{NB} \leq B^{-1}b \text{ (since } x_B \geq 0\text{)}$$

Therefore, any component of x_{NB} can increase only until $\frac{\partial Z}{\partial x_{NB}}$ becomes zero or one or more components of x_B are reduced to zero.

Notes:

1. As the method starts with a basic feasible solution to the constraint equations in $A\underline{x} = \underline{b}$ with $\underline{x} \geq \underline{0}$, the possibility of no solution is ruled out.

2. As the computations proceed, the size of the basis may increase and may decrease to m again.

3. While making a non-basic variable basic if neither any basic variable nor the rate vanishes, then the given QPP will have an unbounded solution.

4. When the termination criterion is satisfied, the corresponding solution will give a local maximum of Z over the feasible region. As $Z(\underline{x})$ is assumed to be concave, this local maximum will also be global.

Example 81 *Use Beale's method to solve the following QPP*

$$\left.\begin{array}{l} \max\ Q(\underline{x}) = 10x_1 + 25x_2 - 10x_1^2 - x_2^2 - 4x_1x_2 \\ \text{subject to } x_1 + 2x_2 \leq 10 \\ \qquad\qquad x_1 + x_2 \leq 9 \\ \text{and } x_1, x_2 \geq 0 \end{array}\right\} \quad (6.25)$$

Using x_3 and x_4 as slack variables, the given constraints can be expressed as:

$$x_1 + 2x_2 + x_3 = 10 \qquad (6.26)$$

$$x_1 + x_2 + x_4 = 9 \qquad (6.27)$$

Iteration 1: We have a starting solution with x_1, x_2 as non basic variables and x_3, x_4 as basic variables:

$$\underline{x}^{(1)} = \begin{pmatrix} x_1 \\ x_2 \\ x_3 \\ x_4 \end{pmatrix} = \begin{pmatrix} 0 \\ 0 \\ 10 \\ 9 \end{pmatrix}$$

Equations (6.26) & (6.27) give the equations of the basic variables in terms of non basic variables as:

$$x_3 = 10 - x_1 - 2x_2 \qquad (6.28)$$

$$x_4 = 9 - x_1 - x_2 \qquad (6.29)$$

The objective function is ready in terms of non basic variables x_1 & x_6. The corresponding value of the objective function is $Q(\underline{x}^{(1)}) = 0$.
Optimality Check:

$$\frac{\partial Q}{\partial x_1} = 10 - 20x_1 - 4x_2 \rightarrow \left(\frac{\partial Q}{\partial x_1}\right)_{\underline{x}^{(1)}} = 10 > 0$$

$$\frac{\partial Q}{\partial x_2} = 25 - 4x_1 - 2x_2 \rightarrow \left(\frac{\partial Q}{\partial x_2}\right)_{\underline{x}^{(1)}} = 25 > 0$$

Hence the optimality criterion is not satisfied. Now, $max\left\{\frac{\partial Q}{\partial x_1}, \frac{\partial Q}{\partial x_2}\right\}_{\underline{x}^{(1)}} = max\{10, 25\} = 25 = \left(\frac{\partial Q}{\partial x_2}\right)_{\underline{x}^{(1)}}$. Therefore, x_2 is to be made basic and x_3 non basic.

Iteration 2: Expressing basic variables x_2, x_4 in terms of non basic variables x_1, x_3 we get:

from (6.28) $\qquad x_2 = 5 - 1/2 x_1 - 1/2 x_3 \qquad (6.30)$

from (6.29) & (6.30) $\qquad x_4 = 4 - 1/2 x_1 + 1/2 x_3 \qquad (6.31)$

Expressing the objective function in terms of the non basic variable we get:

$$Q(x_1, x_3) = 100 - 55/2 x_1 - 15/2 x_3 - 33/4 x_1^2 - 1/4 x_3^2 + 3/2 x_1 x_3 \qquad (6.32)$$

Thus, the new improved solution is given by:

$$\underline{x}^{(2)} = \begin{pmatrix} x_1 \\ x_2 \\ x_3 \\ x_4 \end{pmatrix} = \begin{pmatrix} 0 \\ 5 \\ 0 \\ 4 \end{pmatrix}$$

Optimality Check:

$$\frac{\partial Q}{\partial x_1} = -55/2 - 33/2 x_1 + 3/2 x_3 \rightarrow \left(\frac{\partial Q}{\partial x_1}\right)_{\underline{x}^{(2)}} = -55/2 < 0$$

$$\frac{\partial Q}{\partial x_3} = -15/2 - 1/2x_3 + 3/2x_1 \to \left(\frac{\partial Q}{\partial x_3}\right)_{\underline{x}^{(2)}} = -15/2 < 0$$

Therefore, optimality criterion is satisfied. Thus, the required optimal solution to the given QPP (6.25) is

$$x_1^* = 0, \ x_2^* = 5 \ \text{and} \ Q^* = 100$$

6.6 Solution of Quadratic Programming Problems using LINGO-18

This book focuses on the solution of optimization problems through LINGO-18 Software. And since LINGO-18 follows the standard algorithms for the solution of NLPPs, therefore, some of the QP problems which are solved through Wolfe's and Beale's method are now solved by using LINGO-18 codes in this section.

Example 82 *Solve the following QPP:*

$$\left. \begin{array}{l} \max \ Q(\underline{x}) = 4x_1 + 6x_2 - x_1^2 - 3x_2^2 \\ \text{subject to} \ x_1 + 2x_2 \leq 4 \\ \text{and} \ x_1, x_2 \geq 0 \end{array} \right\} \quad (6.33)$$

The direct LINGO-18 model of the QPP (6.33) is as follows:

LINGO-18 codes

MODEL:
MAX = 4*X1 + 6*X2 − X1^2 − 3*X2^2;
X1 + 2*X2 <= 4;
END

The global optimum solution of the QPP (6.33) is derived using LINGO-18 model as:

$$x_1^* = 2, \ x_2^* = 1 \ \text{with} \ Q^* = 7$$

Example 83 *Solve the following QPP:*

$$\left. \begin{array}{l} \max \ Q(\underline{x}) = 2x_1 + 6x_2 - x_1^2 + 2x_1x_2 - 2x_2^2 \\ \text{subject to} \ x_1 + x_2 \leq 2 \\ -x_1 + 2x_2 \leq 2 \\ \text{and} \ x_1, x_2 \geq 0 \end{array} \right\} \quad (6.34)$$

The direct LINGO-18 model of the QPP (6.34) is as follows:

```
                    LINGO-18 codes
    MODEL :
    MAX = 2*X1 + 6*X2 − X1^2 + 2*X1*X2 − 2*X2^2;
    X1 + X2 <= 2;
    −X1 + 2*X2 <= 2;
    END
```

The global optimum solution of the QPP (6.34) is derived using the LINGO-18 model as:
$$x_1^* = 0.8, \ x_2^* = 1.2 \ \text{with} \ Q^* = 7.2$$

Example 84 *Solve the following QPP:*

$$\left. \begin{array}{l} \max \ Q(\underline{x}) = 10x_1 + 25x_2 - 10x_1^2 - x_2^2 - 4x_1x_2 \\ \text{subject to} \ x_1 + 2x_2 \leq 10 \\ \qquad\qquad x_1 + x_2 \leq 9 \\ \text{and} \ x_1, x_2 \geq 0 \end{array} \right\} \qquad (6.35)$$

The direct LINGO-18 model of the QPP (6.35) is as follows:

```
                    LINGO-18 codes
    MODEL :
    MAX = 10*X1 + 25*X2 − 10*X1^2 − X2^2 − 4*X1*X2;
    X1 + 2*X2 <= 10;
    X1 + X2 <= 9;
    END
```

The global optimum solution of the QPP (6.35) is derived using LINGO-18 model as:
$$x_1^* = 0, \ x_2^* = 5 \ \text{with} \ Q^* = 100$$

Example 85 *A firm produces a single item using two types of raw materials A and B. Material A is available in the market at $100 per unit while B is available at $50 per unit.*

It is known from experience that x_1 units of raw material A produce $2x_1(9 - x_1)$ units of the item while x_2 units of B produces $x_2(8 - x_2)$ units of the item. The firm wants to know how many units of the two raw materials are to be purchased to maximize the production if $500 is available for the purchase of the raw materials. Formulate the above problem as a QPP and solve it by LINGO-18 software.

The formulation of the above problem as a QPP is as follows:

$$\begin{aligned}\max\ Q(\underline{x})\ &= 18x_1 + 8x_2 - 2x_1^2 - x_2^2\\ \text{subject to}\ \ &100x_1 + 50x_2 \le 500\\ \text{or}\ \ &2x_1 + x_2 \le 10\\ \text{and}\ \ &x_1, x_2 \ge 0\end{aligned} \qquad (6.36)$$

The direct LINGO-18 model of the QPP (6.36) is as follows:

LINGO-18 codes

$MODEL:$
$MAX = 18*X1 + 8*X2 - 2*X1^{\wedge}2 - X2^{\wedge}2;$
$2*X1 + X2 <= 10;$
END

The global optimum solution of the QPP (6.36) is derived using LINGO-18 model as:
$$x_1^* = 3.5,\ x_2^* = 3\ \text{with}\ Q^* = 53.5$$

6.7 Convex Programming

The minimization of a convex function or maximization of a concave function subject to a set of convex constraints is termed as a Convex Programming Problem (CPP). The mathematical model for a CPP may be given as:

$$\left.\begin{aligned}\min\ &= f(\underline{x})\\ \text{subject to}\ \ &g_j(\underline{x}) \le 0;\ j = 1, 2, \ldots, m\end{aligned}\right\} \qquad (6.37)$$

where the function $f(\underline{x})$ and $g_j(\underline{x})$; $j = 1, 2, \ldots, m$ are convex.

6.7.1 A Feasible Direction

A direction $\underline{s}$ at a feasible point $\underline{x}$ is called a feasible direction if a small movement from $\underline{x}$ in that direction violates no constraints. Mathematically, $\underline{s}$ will be a feasible direction at a feasible point $\underline{x}$ if $\underline{s}' \nabla g_j(\underline{x}) \le 0$; $j = 1, 2, \ldots, m$.

6.7.2 A Usable Feasible Direction

A feasible direction $\underline{s}$ is said to be usable if a small movement in that direction improves the value of the objective function. Mathematically, a direction $\underline{s}$ will be a usable feasible if $\underline{s}' \nabla g_j(\underline{x}) \le 0$; $j = 1, 2, \ldots, m$ and $\underline{s}' \nabla f(\underline{x}) \le 0$.

6.7.3 An Outline of the Methods of Feasible Directions

These methods are extension of descent methods. Usually a feasible point x_1 is selected as a starting solution. At the ith iteration the current solution x_i is improved by taking the next solution x_{i+1} as $x_{i+1} = x_i + \lambda_i^* s_i$, where λ_i^* denotes the minimizing step length and s_i is a usable feasible direction at x_i. This process of improvement is repeated until no usable feasible direction is available and hence no further improvements in the value of the objective function are possible. The corresponding solution will represent a constrained local minimum of $f(x)$. This local minimum will also be a global minimum because the functions $f(x)$ and $g_j(x)$; $j = 1, 2, \ldots, m$ are convex.

The various methods of feasible directions differ only in the way they select a usable feasible direction. Therefore, in the next sections, two methods to solve CPP are discussed in detail.

6.7.4 Rosen's Gradient Projection Method

This method uses the projection of the negative gradient of the objective function onto the constraints that are currently active to find a usable feasible direction. This method is more efficient for a CPP with linear constraints.

Consider a CPP as:

$$\left. \begin{array}{l} \min \ = f(x) \\ \text{subject to } g_j(x) = \sum_{i=1}^{n} a_{ij} x_i - b_j \leq 0; \ j = 1, 2, \ldots, m \end{array} \right\} \quad (6.38)$$

The Rosen's Gradient Projection algorithm to solve the CPP (6.38) is as follows:

Step 1 Start with an initial feasible solution x_1 by setting the iteration number $i = 1$.

Step 2 If $g_j(x_i) < 0$ for $j = 1, 2, \ldots, m$, set

$$s_i = -\frac{\nabla f(x_i)}{|\nabla f(x_i)|}$$

and go to step 5, otherwise go to step 3.

Step 3 Compute $P_i = I - N_p(N_p' N_p)^{-1} N_p'$ where $N_p = (\nabla g_{j_1}, \nabla g_{j_2}, \ldots, \nabla g_{j_p})$ and $g_j(x_i) = 0$ for $j = j_1, j_2, \ldots, j_p$.

Step 4 Compute

$$s_i = -\frac{P_i \nabla f(x_i)}{|P_i \nabla f(x_i)|}$$

Step 5 If $\underline{s}_i \neq \underline{0}$ go to step 6, otherwise compute $\underline{U} = -(N_p'N_p)^{-1}N_p'\nabla f(\underline{x}_i)$. If $\underline{U} \geq 0$, STOP, $\underline{x}_i = \underline{x}^*$, i.e., solution is optimal, otherwise select the most negative component of $\underline{U}$ say U_q and form N_{p-1} by dropping the qth column of N_p and go to step 3 with $N_p = N_{p-1}$.

Step 6 Compute $\lambda_M = \min_{k(\lambda_k > 0)} \lambda_k$, where k is any integer among $1, 2, \ldots, m$ other than $j_1, j_2, \ldots, j_p$. If

$$\left(\frac{df}{d\lambda}\right)_{\lambda = \lambda_M} = \underline{s}_i' \nabla f(\underline{x}_i + \lambda_M s_i) \leq 0,$$

set $\lambda_i = \lambda_M$, otherwise calculate λ_j^* by some one dimensional unconstrained minimization technique and take $\lambda_i = \lambda_i^*$.

Step 7 Compute $\underline{x}_{i+1} = \underline{x}_i + \lambda_i^* \underline{s}_i$. If some new constraint is active at $\underline{x}_{i+1}$, compute N_{new} by adding one more column to N_p. Set $i = i + 1$ and go to step 3 with $N_p = N_{new}$.
If no new constraint is active at $\underline{x}_{i+1}$, set $i = i + 1$ and go to step 2.

Notes:

(i) If $\lambda_i = \lambda_M$ or if $\lambda_M \leq \lambda_i^*$, one or more new constraints will become active at $\underline{x}_{i+1}$.

(ii) If $\lambda_i = \lambda_i^*$ and $\lambda_i^* < \lambda_M$, no new constraint will become active at $\underline{x}_{i+1}$.

Example 86 *Consider a convex programming problem from [29].*

$$\left.\begin{aligned}
\min \ f(\underline{x}) &= x_1^2 + x_2^2 - 2x_1 - 4x_2 \\
\text{subject to} \ \ x_1 + 4x_2 - 5 &\leq 0 \\
2x_1 + 3x_2 - 6 &\leq 0 \\
-x_1 &\leq 0 \\
-x_2 &\leq 0
\end{aligned}\right\} \quad (6.39)$$

We start from the point

$$X_1 = \begin{pmatrix} 1 \\ 1 \end{pmatrix}$$

Iteration i = 1:
Since $g_j(X_1) = 0$ for $j = 1$, we have $p = 1$ & $j_1 = 1$. Now, following step 4 of the algorithm, we have

$$N_1 = \begin{bmatrix} 1 \\ 4 \end{bmatrix}$$

$$P_1 = \frac{1}{17}\begin{bmatrix} 16 & -4 \\ -4 & 1 \end{bmatrix}$$

$$\nabla f(X_1) = \begin{bmatrix} 2x_1 - 2 \\ 2x_2 - 4 \end{bmatrix}_{X_1} = \begin{bmatrix} 0 \\ -2 \end{bmatrix}$$

and

$$S_1 = \begin{bmatrix} -0.4707 \\ 0.1177 \end{bmatrix}$$

Now, the normalized search direction can be obtained as:

$$S_1 = \frac{1}{\sqrt{(-0.4707)^2 + (0.1177)^2}} \times \begin{bmatrix} -0.4707 \\ 0.1177 \end{bmatrix} = \begin{bmatrix} -0.9701 \\ 0.2425 \end{bmatrix}$$

Since $S_1 \neq 0$, we go to step 6 and find the step length λ_M as:

$$X = \begin{bmatrix} x_1 \\ x_2 \end{bmatrix} = X_1 + \lambda S_1 = \begin{bmatrix} 1 - 0.9701\lambda \\ 1 + 0.2425\lambda \end{bmatrix}$$

Therefore, $\lambda_M = \lambda_2 = 1.03$. Also,

Table 6.4: Constraint Value at Different j.

j	$g_j(X)$	Value at λ_j
2	$(2-1.9402\lambda)+(3+0.7275\lambda)-6=0$	$\lambda_2 = -0.8245$
3	$-(1-0.9401\lambda)=0$	$\lambda_3 = 1.03$
4	$-(1+0.2425\lambda)=0$	$\lambda_4 = -4.124$

$$f(X) = f(\lambda) = 0.9998\lambda^2 - 0.4850\lambda - 4$$

$$\frac{df}{d\lambda} = 1.9996\lambda - 0.4850$$

$$\left(\frac{df}{d\lambda}\right)_{\lambda_M} = 1.9996(1.03) - 0.4850 = 1.5746 > 0$$

$$\Rightarrow \frac{df}{d\lambda} = 0 = 0.2425$$

Now, obtain new point X_2 as:

$$X_2 = X_1 + \lambda_1 S_1 = \begin{bmatrix} 0.7647 \\ 1.0588 \end{bmatrix}$$

Since $\lambda = \lambda_1^*$ and $\lambda_1^* < \lambda_{M^*}$ no new constraint has become active at X_2 and hence the matrix N_1 remains unaltered.

Iteration 2: i = 2
$g_j(X_2) = 0$ for $j = 1$, we have $p = 1$, $f_1 = 1$ and go to step 4 of the algorithm, we have

$$N_1 = \begin{bmatrix} 1 \\ 4 \end{bmatrix}$$

$$P_2 = \frac{1}{17} \begin{bmatrix} 16 & -4 \\ -4 & 1 \end{bmatrix}$$

$$\nabla f(X_2) = \begin{bmatrix} 2x_1 - 2 \\ 2x_2 - 4 \end{bmatrix}_{X_2} = \begin{bmatrix} -0.4706 \\ -1.8824 \end{bmatrix}$$

and

$$S_2 = -P_2 \nabla f(X_2) = \begin{bmatrix} 0 \\ 0 \end{bmatrix}$$

Since $S_2 = 0$, we compute the vector λ at X_2 as

$$\lambda = -(N_1^T N_1)^{-1} N_1^T \nabla f(X_2) = 0.4707 > 0$$

The non-negative value of λ indicates that we have reached the optimum point and hence

$$X_{opt} = X_2 = \begin{bmatrix} 0.7647 \\ 1.0588 \end{bmatrix} \text{ with } f_{opt} = -4.059$$

6.7.5 Kelly's Method

Consider the CPP:

$$\left. \begin{array}{l} \min \ = f(\underline{x}) \\ \text{subject to } g_j(\underline{x}) \leq 0; \ j = 1, 2, \ldots, m \end{array} \right\} \quad (6.40)$$

Without loss of generality we can assume that $f(\underline{x})$ is linear. If $f(\underline{x})$ is not linear we can rewrite the CPP as:

$$\left. \begin{array}{l} \min \ = x_{n+1} \\ \text{subject to } g_j(\underline{x}) \leq 0; \ j = 1, 2, \ldots, m \\ \text{and } x_{n+1} - f(\underline{x}) = 0 \end{array} \right\} \quad (6.41)$$

Thus, without loss of generality we can write the CPP as:

$$\left. \begin{array}{l} \min \ = \underline{c}' \underline{x} \\ \text{subject to } g_j(\underline{x}) \leq 0; \ j = 1, 2, \ldots, m \end{array} \right\} \quad (6.42)$$

where some or all constraints may be non-linear.

The stepwise algorithm of the Kelly's method is given below:

Step 1 Start with an initial point $\underline{x}_1$ and set the iteration number $i = 1$. The point $\underline{x}_1$ need not be feasible.

Step 2 Linearize the non-linear constraints $g_j(\underline{x})$ about $\underline{x}_i$ as

$$g_j(\underline{x}) \cong g_j(\underline{x}_i) + \nabla' g_j(\underline{x}_i)(\underline{x} - \underline{x}_i)$$

Step 3 Construct the approximated LPP as:

$$\left.\begin{array}{l} \min \ = \underline{c}'\underline{x} \\ \text{subject to } \ g_j(\underline{x}_i) + \nabla' g_j(\underline{x}_i)(\underline{x} - \underline{x}_i) \leq 0; \ \ j = 1, 2, \ldots, m \end{array}\right\} \quad (6.43)$$

Step 4 Solve the LPP. Let $\underline{x}_{i+1}$ be the optimal solution.

Step 5 Compute $g_j(\underline{x}_{i+1})$; $j = 1, 2, \ldots, m$. If $g_j(\underline{x}_{i+1}) \leq \varepsilon$; $j = 1, 2, \ldots, m$, STOP. $\underline{x}_{i+1} = \underline{x}^*$ is the required optimal solution to the original CPP. Otherwise go to step 6.

Step 6 Linearize the kth constraint $g_k(\underline{x})$ about $\underline{x}_{i+1}$ and add it to the set of constraints of the LPP constructed at step 3, where the kth constraint is the most violated constraint, that is

$$g_k(\underline{x}_{i+1}) = \max\{g_j(\underline{x}_{i+1}) | g_j(\underline{x}_{i+1}) > \varepsilon\}$$

Step 7 Set $i = i + 1$ and $m = m + 1$ and go to step 4.

Notes:

(i) $\varepsilon > 0$ is a preassigned tolerance limit fixed according to the required precision of the approximate optimal solution to the original CPP (6.40).

(ii) This method is called 'Cutting Plane Method' because the additional linear constraints used for linearizing the original constraints about various optimal points of the approximated LPP at each iteration are hyperplane a portion of the existing feasible region is cut off.

Example 87 *Consider a convex programming problem from [29].*

$$\left.\begin{array}{l} \min \ f(\underline{x}) = x_1 - x_2 \\ \text{subject to } \ 3x_1^2 - 2x_1 x_2 + x_2^2 - 1 \leq 0 \end{array}\right\} \quad (6.44)$$

Following steps 1, 2, & 3, before starting with any initial point X_1 to avoid the possible unbounded solution, we first consider the bounds on x_1 & x_2 as $-2 \leq x_1 \leq 2$ and $-2 \leq x_2 \leq 2$, and solve the following LPP:

$$\left.\begin{array}{l} \min \ f(\underline{x}) = x_1 - x_2 \\ \text{subject to} \ -2 \leq x_1 \leq 2 \\ \phantom{\text{subject to}} \ -2 \leq x_2 \leq 2 \end{array}\right\} \quad (6.45)$$

The solution of this problem can be obtained as:

$$X = \begin{bmatrix} -2 \\ 2 \end{bmatrix} \text{ with } f(X) = -4$$

Step 4: Since we have solved one LPP, we can take

$$X_{i+1} = X_2 = \begin{bmatrix} -2 \\ 2 \end{bmatrix}$$

Step 5: Since $g_1(X_2) = 23 > \varepsilon$, we linearize $g_1(X)$ about the point X_2 as

$$g_1(X) \simeq g_1(X_2) + \nabla g_1(X_2)^T (X - X_2) \leq 0 \quad (6.46)$$

As

$$g_1(X_2) = 23, \ \frac{\partial g_1}{\partial x_1}\Big|_{X_2} = (6x_1 - 2x_2)|_{X_2} = -16$$

and

$$\frac{\partial g_1}{\partial x_2}\Big|_{X_2} = (-2x_1 + 2x_2)|_{X_2} = 8$$

(6.46) becomes

$$g_1(X) \simeq -16x_1 + 8x_2 - 25 \leq 0$$

By adding this constraint to the previous LPP, the new LPP becomes:

$$\left.\begin{array}{l} \min \ f(\underline{x}) = x_1 - x_2 \\ \text{subject to} \ -2 \leq x_1 \leq 2 \\ \phantom{\text{subject to}} \ -2 \leq x_2 \leq 2 \\ \phantom{\text{subject to}} \ -16x_1 + 8x_2 - 25 \leq 0 \end{array}\right\} \quad (6.47)$$

Now setting the iteration number be i=2 and moving to step 4. Solve the approximating LPP stated in (6.47) and obtain the solution:

$$X_3 = \begin{bmatrix} -0.5625 \\ 2 \end{bmatrix} \text{ with } f_2 = f(X_3) = -2.5625$$

As $g_1(X_3) = 6.19922 > \varepsilon$, we linearize $g_1(X)$ about X_3 as given in Step 6.
Step 6: Since $g_1(X_2) = 23 > \varepsilon$, we linearize $g_1(X)$ about the point X_2 as

$$g_1(X) \simeq g_1(X_3) + \nabla g_1(X_3)^T (X - X_3) \leq 0 \quad (6.48)$$

As

$$g_1(X_3) = 6.19922, \quad \frac{\partial g_1}{\partial x_1}\bigg|_{X_3} = -7.375$$

and

$$\frac{\partial g_1}{\partial x_2}\bigg|_{X_3} = 5.125$$

(6.48) becomes

$$g_1(X) \simeq -7.375x_1 + 5.125x_2 - 8.19922 \leq 0$$

By adding this constraint to the previous LPP, the new LPP becomes:

$$\left.\begin{array}{l} \min \ f(\underline{x}) = x_1 - x_2 \\ \text{subject to} \ -2 \leq x_1 \leq 2 \\ \phantom{\text{subject to}} \ -2 \leq x_2 \leq 2 \\ \phantom{\text{subject to}} \ -16x_1 + 8x_2 - 25 \leq 0 \\ \phantom{\text{subject to}} \ -7.375x_1 + 5.125x_2 - 8.19922 \leq 0 \end{array}\right\} \quad (6.49)$$

Now set the iteration number as i=3 and go to step 4. Solve the approximating LPP (6.49) to obtain the solution:

$$X_4 = \begin{bmatrix} 0.27870 \\ 2 \end{bmatrix} \quad \text{with} \quad f_4 = f(X_4) = -1.72193$$

This procedure is continued until the specified convergence criterion, $g_j(X_i) \leq \varepsilon$ in step 5 is satisfied. The condition satisfied at the tenth iteration, and the result is as follows:

$$X = \begin{bmatrix} 0.000109 \\ 1 \end{bmatrix} \quad \text{with} \quad f_{opt} = -1$$

6.8 Solution of Convex Programming Problems using LINGO-18

In this section, the same problems which were solved in previous sections through algorithms are solved utilizing LINGO-18 software and merely derived the same results. Some other convex programming problems are also solved through LINGO-18 software.

Example 88 *Consider a convex programming problem from [29].*

$$\left.\begin{array}{l} \min \ f(\underline{x}) = x_1^2 + x_2^2 - 2x_1 - 4x_2 \\ \text{subject to} \ x_1 + 4x_2 - 5 \leq 0 \\ \phantom{\text{subject to}} \ 2x_1 + 3x_2 - 6 \leq 0 \\ \phantom{\text{subject to}} \ -x_1 \leq 0 \\ \phantom{\text{subject to}} \ -x_2 \leq 0 \end{array}\right\} \quad (6.50)$$

The direct LINGO-18 model of the CPP (6.50) is as follows:

LINGO-18 codes
MODEL: MIN= X1 ^ 2 + X2 ^ 2 - 2 * X1 - 4 * X2; X1 + 4 * X2 <= 5; 2 * X1 + 3 * X2 <= 6; - X1 <= 0; - X2 <= 0; END

The global optimum solution of the CPP (6.50) is derived using the LINGO-18 model as:

$$x_1^* = 0.7647,\ x_2^* = 1.0588 \text{ with } f_{opt} = -4.059$$

Example 89 Consider a convex programming problem from [29].

$$\left.\begin{array}{l} \min\ f(\underline{x}) = x_1 - x_2 \\ \text{subject to}\ 3x_1^2 - 2x_1x_2 + x_2^2 - 1 \leq 0 \end{array}\right\} \quad (6.51)$$

The direct LINGO-18 model of the CPP (6.51) is as follows:

LINGO-18 codes
MODEL: MIN= X1 - X2; 3 * X1 ^ 2 - 2 * X1 * X2 + X2 ^ 2 - 1 <= 0; END

The global optimum solution of the CPP (6.51) is derived using the LINGO-18 model as:

$$x_1^* = 0.000109,\ x_2^* = 1 \text{ with } f_{opt} = -1$$

Example 90 Consider the following convex programming problem and derive the optimum solution using LINGO-18 software.

$$\left.\begin{array}{l} \min\ f(\underline{x}) = 2x_1^2 + 2x_2^2 - 2x_1x_2 - 4x_1 - 6x_2 \\ \text{subject to}\ x_1 + x_2 \leq 2 \\ \qquad\qquad x_1 + 5x_2 \leq 5 \\ \qquad\qquad -x_1 \leq 0 \\ \qquad\qquad -x_2 \leq 0 \end{array}\right\} \quad (6.52)$$

The direct LINGO-18 model of the CPP (6.52) is as follows:

LINGO-18 codes

MODEL:
MIN= 2 * X1 ^ 2 + 2 * X2 ^ 2 - 2 * X1 * X2 - 4 * X1 - 6 * X2;
X1 + X2 <= 2;
X1 + 5 * X2 <= 5;
- X1 <= 0;
- X2 <= 0;
END

The global optimum solution of the CPP (6.52) is derived using LINGO-18 model as:

$$x_1^* = 1.129108, \; x_2^* = 0.7741784 \; \text{with} \; f_{opt} = -7.161290$$

Example 91 Consider the following convex programming problem and derive the optimum solution using LINGO-18 software.

$$\left. \begin{array}{l} \min \; f(\underline{x}) = x_1^2 + 2x_2^2 + 3x_3^2 + x_1x_2 - 2x_1x_3 + x_2x_3 - 4x_1 - 6x_2 \\ \text{subject to} \; x_1 + 2x_2 + x_3 \leq 4 \\ \phantom{\text{subject to}} \; x_1, x_2, x_3 \leq 0 \end{array} \right\} \quad (6.53)$$

The direct LINGO-18 model of the CPP (6.53) is as follows:

LINGO-18 codes

MODEL: MIN= X1 ^ 2 + 2 * X2 ^ 2 + 3 * X3 ^ 2 + X1 * X2 - 2 * X1 * X3 + X2 * X3 - 4 * X1 - 6 * X2;
X1 + 2 * X2 + X3 <= 4;
END

The global optimum solution of the CPP (6.53) is derived using LINGO-18 model as:

$$x_1^* = 1.999999, \; x_2^* = 0.7499998, \; x_3^* = 0.5000010 \; \text{with} \; f_{opt} = -6.750000$$

Chapter 7
Optimization Under Uncertainty

7.1 Introduction

In real-life optimization problems where the data values or parameters either in the objective function or in constraints or in both are not certainly known are popularly known as Uncertain Optimization Problems. Uncertainty may be in the form of vague values, random values or interval values. This chapter dealt with the optimization problems under such kind of uncertainty. The different types of problems depending upon the nature of uncertainty are fuzzy optimization problems used when the parameters are vaguely defined, stochastic optimization problem when the parameters are random in nature and interval optimization problem when the parameters are given in interval form. In the following sections, concepts of fuzzy, stochastic, and interval have been briefly discussed.

7.2 Fuzzy Optimization

7.2.1 Introduction

Fuzzy logic is an extension of Boolean logic which was developed by [36], based on mathematical theory of fuzzy sets, which is a generalization of the classical set theory. In a classical set theory, an element of a set either belongs or does not belong to the set. In fuzzy set theory, an element belongs to the set with a membership grade in an interval of [0,1]. All membership grades together form the membership function. A classical set is often called crisp as opposed to fuzzy.

The membership of a crisp set A is defined by

$$\mu_A(x) = \begin{cases} 0, & if\ x \in A \\ 1, & if\ x \notin A \end{cases} \quad (7.1)$$

Note that, the membership function of a crisp set can take a value of either 0 or 1 while the fuzzy membership functions can take any value in the closed interval [0,1]. Further, some basic definitions and concepts of fuzzy set theory are summarized from [13, 35, 18, 21].

Remarks: The classical set theory is a subset of the theory of fuzzy sets.

Definition 7.1 A set is a collection of elements a_0 or members. A set may be an element of another set.

Definition 7.2 Let X be a set of elements x. A fuzzy set A is a collection of ordered pairs $(x, \mu_A(x))$ for $x \in X$. X is called the universe of discourse and $\mu_A(x) : \longrightarrow [0,1]$ is the membership function. The function $\mu_A(x)$ provides the degree of fulfilment of $x \in X$. When X is countable, the fuzzy set is represented as

$$A = \frac{\mu_A(x_1)}{x_1} \oplus \frac{\mu_A(x_2)}{x_2} \oplus \ldots \oplus \frac{\mu_A(x_n)}{x_n} \quad (7.2)$$

The positive sign $\oplus$ is a common notation in the context of fuzzy which states the element x_i of X and the corresponding membership grades. For example, "temperature" is a linguistic variable whose values can be very high; moderate; low; very low. Each of these values can be represented as a fuzzy set.

Example 92 *Consider a patient temperature in degree Celsius as $X = \{46, 46.5, 47, 47.5, \ldots, 50.5\}$. Let the fuzzy set $A =$ "high temperature", and defined by*

$$A = \{\mu_A(x)/x | x \in X\} \quad (7.3)$$

$$= \frac{0}{46} \oplus \frac{0}{46.5} \oplus \frac{0.1}{47} \oplus \frac{0.3}{47.5} \oplus \frac{0.7}{48} \oplus \frac{0.8}{48.5} \oplus \frac{0.8}{49} \oplus \frac{0.9}{49.5} \oplus \frac{1}{50} \oplus \frac{1}{50.5},$$

where the numbers $0, 0.1, 0.3, 0.7, 0.8, 0.9, \& 1$ express the degree to which the corresponding temperature is high.

Definition 7.3 The *support* of a fuzzy set A is the crisp set of all elements of X with nonzero membership in A. Symbolically,

$$supp(A) = \{x \in X | \mu_A(x) > 0\} \quad (7.4)$$

Example 93 *Using example (92), $supp(A) = \{47, 47.5, 48, \ldots, 50.5\}$.*

Definition 7.4 The *height* of A, denoted by $h(A)$, corresponds to the upper bound of the codomain of its membership function. That is, the *height* of a fuzzy set A on X is defined as

$$h(A) = sup_{x \in X} \{\mu_A(x) | x \in X\} \tag{7.5}$$

Definition 7.5 A fuzzy set A is said to be *normal* $\iff h(A) = 1$. Otherwise it is a *subnormal*.

Definition 7.6 The nucleus of a fuzzy set A is the set of values x for which $\mu_A(x) = 1$.

Definition 7.7 The *kernel* of (A) is the set of elements of X belonging entirely to A. In other words, the kernel

$$noy(A) = \{x \in X | \mu_A(x) = 1\} \tag{7.6}$$

By construction, $noy(A) \subseteq supp(A)$.

Definition 7.8 The set of all elements of X with membership in A at least α is called the $\alpha - level\ set$. That is,

$$A_\alpha = \{x \in X | \mu_A(x) \geq \alpha\} \tag{7.7}$$

Definition 7.9 A fuzzy set A is said to be *convex* if the membership function is *quasi concave*; that is, $\forall x_1, x_2 \in X$, and $\lambda \in [0, 1]$, the following is true:

$$\mu_A \{\lambda x_1 + (1 - \lambda)x_2\} \geq \min \{\mu_A(x_1), \mu_A(x_2)\}. \tag{7.8}$$

Definition 7.10 A fuzzy number A is a fuzzy set in the real R for which the following are true:

i A is normal ($\exists x: \mu_A(x) = 1$).

ii A is convex.

iii A is upper semi continuous.

iv A has bounded support.

7.2.2 Operations of Fuzzy Sets

The basic assumptions, theory, and notations concerning operations on crisp sets will now be extended to the fuzzy sets.

Definition 7.11 Two fuzzy sets A and B in X are equal if:

$$\mu_A(x) = \mu_B(x), \forall\ x \in X.\ i.e\ A = B. \tag{7.9}$$

Definition 7.12 A fuzzy set A in X is a subset of another fuzzy set B also in X if:

$$\mu_A(x) \leq \mu_B(x), \forall\ x \in X. \tag{7.10}$$

Definition 7.13 The following membership functions are defined:
  i Complement $\bar{A}$ of a fuzzy set A in X. $\mu_{\bar{A}} = 1 - \mu_A(x), \forall\ x \in X$.
  ii Union $A \bigcup B$ of two fuzzy sets in X. $\mu_{A \bigcup B} = \max[\mu_A(x), \mu_B(x)], \forall\ x \in X$.
  iii Intersection $A \bigcap B$ of two fuzzy sets in X. $\mu_{A \bigcap B} = \min[\mu_A(x), \mu_B(x)], \forall\ x \in X$.

Example 94 *In continuity of example (92), lets define a new fuzzy set B = "Dangerous temperature" as* $B = \left\{ \dfrac{0}{47.5}, \dfrac{0.1}{48}, \dfrac{0.2}{48.5}, \dfrac{0.5}{49}, \dfrac{0.8}{49.5}, \dfrac{0.9}{50}, \dfrac{1}{50.5} \right\}$. *According to definition (7.13), we have*
$A \bigcup B$ = *"High or dangerous temperature"*
$= 0/46.5 \oplus 0/47 \oplus 0.1/47.5 \oplus 0.5/48 \oplus 0.8/48.5 \oplus 1/49 \oplus 1/49.5 \oplus 1/50 \oplus 1/50.5$
$\bar{A}$ = *"Not high temp"*
$= 1/46.5 \oplus 1/47 \oplus 0.9/47.5 \oplus 0.5/48 \oplus 0.2/48.5 \oplus 0/49 \oplus 0/49.5 \oplus 0/50$

Example 95 *Let fuzzy set $A = \{x, \mu_A(x) : x \in X\}$, where $\mu_A(x) = \dfrac{1}{x^2+1}$. Find the membership value for x equals to $1, 2$, and 3 in fuzzy set A.*

We have $\mu_A(x) = \dfrac{1}{x^2+1}$, then
$\mu_A(1) = \dfrac{1}{1+1} = 0.5$, $\mu_A(2) = \dfrac{1}{2^2+1} = 0.2$ and $\mu_A(3) = \dfrac{1}{3^2+1} = 0.1$

Example 96 *Let fuzzy set A on the universal set $= \{1, 2, \ldots, 5\}$, where $\mu_A(x) = \dfrac{1}{1+x}$, then $\mu_A(1) = 0.5$, $\mu_A(2) = 0.33$ $\mu_A(3) = 0.25$, $\mu_A(4) = 0.2$, and $\mu_A(5) = 0.16$. Thus, fuzzy set $A = \{(1, 0.5), (2, 0.33), (3, 0.25)(4, 0.2), (5, 0.16)\}$.*

Example 97 *Let fuzzy set $A = \{x, \mu_A(x) : x \in X\}$,*

$$\mu_A(x) = \begin{cases} 0, & \text{if } 0 \leq x \leq 4 \\ \dfrac{x-4}{3}, & \text{if } 4 < x < 7 \\ 1, & \text{if } 7 \leq x \leq 11 \\ \dfrac{13-x}{2}, & \text{if } 11 < x \leq 13 \\ 0, & \text{if } x \geq 13 \end{cases} \tag{7.11}$$

The membership at $x = 6$ and $x = 11$ are: $\mu_A(6) = \dfrac{x-4}{3} = 0.67$ *and* $\mu_A(11) = \dfrac{13-x}{2} = 1$.

7.2.3 Cardinality of Fuzzy Set

There are two cardinalities of fuzzy sets.

7.2.3.1 Scalar Cardinality

The cardinality of fuzzy set A is the sum of the membership degree for all $x \in X$, where X is the universal set from which the elements of A are taken. That is,

$$|A| = \sum_{x \in X} \mu_A(x), \quad for\ discrete\ case$$
$$|A| = \int_{x \in X} \mu_A(x), \quad for\ continuous\ case \quad (7.12)$$

7.2.3.2 Relative Cardinality

It is the ratio of the cardinality of A and the universal set X and denoted by $||A||$, where

$$||A|| = \frac{|A|}{|X|} \quad (7.13)$$

Example 98 *Let fuzzy set*
$A = \{(1,0.32),(2,0.4),(3,0.53)(4,0.67),(5,1),(6,1),(7,0.5),(8,0.2),(9,0.1)\}$.
Hence, cardinality of fuzzy set A is $|A| = \sum_{x \in X} \mu_A(x) = 4.72$

Example 99 *Let fuzzy set* $A = \{x, \mu_A(x) : x \in X\}$,

$$\mu_A(x) = \begin{cases} 0, & if\ 0 \leq x \leq 3 \\ \frac{x-3}{3}, & if\ 3 < x < 6 \\ 1, & if\ 6 \leq x \leq 10 \\ \frac{12-x}{2}, & if\ 10 < x \leq 12 \\ 0, & if\ x \geq 12 \end{cases} \quad (7.14)$$

*Here, $\mu_A(x)$ is continuous, thus cardinality of fuzzy set A is $|A| = \int_{x \in X} \mu_A(x)$
Thus*

$$|A| = \int_3^6 \frac{x-3}{3}dx + \int_6^{10} dx + \int_{10}^{12} \frac{12-x}{2}dx = 6.5 \quad (7.15)$$

Example 100 *Let fuzzy set A and its membership function is*

$$\mu_A(x) = \begin{cases} 0, & if\ 0 \leq x \leq 1 \\ x-1, & if\ 1 \leq x < 2 \\ 1, & if\ 2 \leq x \leq 3 \\ 4-x & if\ 3 < x < 4 \\ 0, & if\ x \geq 4 \end{cases} \quad (7.16)$$

Cardinality of fuzzy set A is $|A| = \int_{x \in X} \mu_A(x)$
Thus

$$|A| = \int_1^2 (x-1)dx + \int_2^3 dx + \int_3^4 (12-x)dx = 2 \qquad (7.17)$$

Example 101   i. Consider a fuzzy set A whose membership function is

$$\mu_A(x) = \begin{cases} 0, & if\ 0 \leq x \leq 1 \\ x-1, & if\ 1 \leq x < 2 \\ 1, & if\ 2 \leq x \leq 3 \\ 4-x & if\ 3 < x < 4 \\ 0, & if\ x \geq 4 \end{cases} \qquad (7.18)$$

Is the given set A a normal?
Yes, since $\mu_A(x) = 1$, for $2 \leq x \leq 3$. So there exists $x \in X$ such that $\mu_A(x) = 1$. In light of definition 5, we can say set A is normal.

ii. Consider another fuzzy set B whose membership function is

$$\mu_B(x) = \begin{cases} 0, & if\ 0 \leq x \leq 1 \\ (x-1)0.8, & if\ 1 \leq x < 2 \\ 0.8, & if\ 2 \leq x \leq 3 \\ 0.8(4-x) & if\ 3 < x < 4 \\ 0, & if\ x \geq 4 \end{cases} \qquad (7.19)$$

Since there exists no $x \in X$ such that $\mu_B(x) = 1$, then set B is a subnormal fuzzy set.

Example 102 *Consider two fuzzy sets A and B on universal set* $X = \{1, 2, \ldots, 6\}$, *where* $A = \left\{ \frac{0.3}{1} \oplus \frac{1}{2} \oplus \frac{0.5}{3} \oplus \frac{0.6}{4} \oplus \frac{0.2}{5} \oplus \frac{0.6}{6} \right\}$,
$B = \left\{ \frac{0.9}{1} \oplus \frac{0.5}{2} \oplus \frac{0.8}{3} \oplus \frac{0.4}{4} \oplus \frac{0.7}{5} \oplus \frac{0.3}{6} \right\}.$

Firstly, we define membership function of the standard complement of A as

$$\mu_{\bar{A}} = 1 - \mu_A(x), \forall x \in X.$$

$\mu_{\bar{A}}(1) = 0.7$, $\mu_{\bar{A}}(2) = 0$, *and so on.* $\mu_{\bar{B}}(1) = 0.1$, $\mu_{\bar{B}}(2) = 0.5$, *and so on.*

Now we define
$\mu_{\bar{A}} = \left\{ \frac{0.7}{1} \oplus \frac{0}{2} \oplus \frac{0.5}{3} \oplus \frac{0.4}{4} \oplus \frac{0.8}{5} \oplus \frac{0.4}{6} \right\}$, $\mu_{\bar{B}} = \left\{ \frac{0.1}{1} \oplus \frac{0.5}{2} \oplus \frac{0.2}{3} \oplus \frac{0.6}{4} \oplus \frac{0.3}{5} \oplus \frac{0.7}{6} \right\}.$
$\mu_{A \cup B}(x) = \max\{\mu_A(x), \mu_B(x)\}, x \in X.$ $\mu_{A \cup B}(1) = \max\{\mu_A(1), \mu_B(1)\} = \max\{0.3, 0.9\} = 0.9.$ $\mu_{A \cup B}(2) = \max\{\mu_A(2), \mu_B(2)\} = \max\{1, 0.5\} = 1,$ *and*

so on. Hence $\mu_{A \cup B} = \left\{ \dfrac{0.9}{1} \oplus \dfrac{1.0}{2} \oplus \dfrac{0.8}{3} \oplus \dfrac{0.6}{4} \oplus \dfrac{0.7}{5} \oplus \dfrac{0.6}{6} \right\}$. Similarly, membership function of standard intersection of A and B is given by $\mu_{A \cap B}(x) = \min\{\mu_A(x), \mu_B(x)\}, \forall x \in X$. $\mu_{A \cap B}(1) = \min\{\mu_A(1), \mu_B(1)\} = \min\{0.3, 0.9\} = 0.3$.
$\mu_{A \cap B}(2) = \min\{\mu_A(2), \mu_B(2)\} = \min\{1, 0.5\} = 0.5$, and so on. Therefore,
$\mu_{A \cap B} = \left\{ \dfrac{0.3}{1} \oplus \dfrac{0.5}{2} \oplus \dfrac{0.5}{3} \oplus \dfrac{0.4}{4} \oplus \dfrac{0.2}{5} \oplus \dfrac{0.3}{6} \right\}$.

Example 103 *Let fuzzy set A and its membership function be given as follows:*

$$\mu_A(x) = \begin{cases} 0, & if\ 0 \leq x \leq 3 \\ \dfrac{x-3}{3}, & if\ 3 < x < 6 \\ 1, & if\ 6 \leq x \leq 10 \\ \dfrac{12-x}{2}, & if\ 10 < x \leq 12 \\ 0, & if\ x \geq 12 \end{cases} \qquad (7.20)$$

the corresponding $\mu_{\bar{A}} = 1 - \mu_A(x), \forall x \in X$ *is*

$$\mu_{\bar{A}}(x) = \begin{cases} 1, & if\ 0 \leq x \leq 3 \\ 1 - \dfrac{x-3}{3}, & if\ 3 < x < 6 \\ 0, & if\ 6 \leq x \leq 10 \\ 1 - \dfrac{12-x}{2}, & if\ 10 < x \leq 12 \\ 1, & if\ x \geq 12 \end{cases} \qquad (7.21)$$

Example 104 *Let the membership function of fuzzy set A, and B be given by*

$$\mu_A(x) = \begin{cases} 1, & if\ 0 \leq x \leq 3 \\ \dfrac{6-x}{3}, & if\ 3 < x < 6 \\ 0, & if\ x \geq 6 \end{cases} \qquad (7.22)$$

and

$$\mu_B(x) = \begin{cases} 0, & if\ 0 \leq x \leq 3 \\ \dfrac{x-3}{3}, & if\ 3 < x < 6 \\ 1, & if\ 6 \leq x \leq 10 \\ \dfrac{12-x}{2}, & if\ 10 < x \leq 12 \\ 0, & if\ x \geq 12 \end{cases} \qquad (7.23)$$

By definition we know that

$$\mu_{A \cup B}(x) = \max\{\mu_A(x), \mu_B(x)\}, x \in X.$$

$$= \begin{cases} 1, & \text{if } 0 \leq x \leq 3 \\ \max\left[\dfrac{6-x}{3}, \dfrac{x-3}{3}\right] & \text{if } 3 < x < 6 \\ 1, & \text{if } 6 \leq x \leq 10 \\ \dfrac{12-x}{2}, & \text{if } 10 < x \leq 12 \\ 0, & \text{if } x \geq 12 \end{cases} \tag{7.24}$$

and
$$\mu_{A \cap B}(x) = \min\{\mu_A(x), \mu_B(x)\}, \forall x \in X.$$

$$= \begin{cases} 0, & \text{if } 0 \leq x \leq 3 \\ \min\left[\dfrac{6-x}{3}, \dfrac{x-3}{3}\right], & \text{if } 3 < x < 6 \\ 0, & \text{if } x \geq 6 \end{cases} \tag{7.25}$$

Note that: Fuzzy sets on universal set X satisfied all properties of the crisp set, but the following properties of crisp sets are not satisfied for all fuzzy sets. That is,

i Law of contradiction: $A \cap \bar{A} = \Phi$ i.e., $\min\{\mu_A(x), \mu_{\bar{A}}(x)\} = 0, \forall x \in X$.

ii Law of excluded middle: $A \cup \bar{A} = X$ i.e., $\max\{\mu_A(x), \mu_{\bar{A}}(x)\} = 1, \forall x \in X$.

Example 105 *Consider fuzzy sets A on universal set $X = \{1,2,3,4\}$, where*
$A = \left\{\dfrac{0.1}{1} \oplus \dfrac{0.7}{2} \oplus \dfrac{0.5}{3} \oplus \dfrac{01.0}{4}\right\}$, $\bar{A} = \left\{\dfrac{0.9}{1} \oplus \dfrac{0.3}{2} \oplus \dfrac{0.5}{3} \oplus \dfrac{0.0}{4}\right\}$. *By definition of law of contradiction $A \cap \bar{A} = \Phi$ i.e., $\min\{\mu_A(x), \mu_{\bar{A}}(x)\} = 0, \forall x \in X$.*

$\min\{\mu_A(1), \mu_{\bar{A}}(1)\} = \min\{0.1, 0.9\} = 0.1$,
$\min\{\mu_A(2), \mu_{\bar{A}}(2)\} = \min\{0.7, 0.3\} = 0.3$,
$\min\{\mu_A(3), \mu_{\bar{A}}(3)\} = \min\{0.5, 0.5\} = 0.5$.
$\min\{\mu_A(4), \mu_{\bar{A}}(4)\} = \min\{1.0, 0.0\} = 0.0$,
Hence, $\min\{\mu_A(x), \mu_{\bar{A}}(x)\} \neq 0, \forall x \in X$, *and law of contradiction does not hold.*

By the definition of Law of excluded middle:
$A \cup \bar{A} = X$ i.e., $\max\{\mu_A(x), \mu_{\bar{A}}(x)\} = 1, \forall x \in X$.
$\max\{\mu_A(1), \mu_{\bar{A}}(1)\} = \max\{0.1, 0.9\} = 0.9$,
$\max\{\mu_A(2), \mu_{\bar{A}}(2)\} = \max\{0.7, 0.3\} = 0.7$,
$\max\{\mu_A(3), \mu_{\bar{A}}(3)\} = \max\{0.5, 0.5\} = 0.5$.
$\max\{\mu_A(4), \mu_{\bar{A}}(4)\} = \max\{1.0, 0.0\} = 1.0$,
Hence, $\max\{\mu_A(x), \mu_{\bar{A}}(x)\} \neq 1, \forall x \in X$, *law of excluded middle does not hold.*

7.2.3.3 α-cut Set

In 1965, [36] introduced the concept of α-cut set to establish a bridge between fuzzy set theory and crisp set theory. α-cut set of the fuzzy set A is the crisp set that contains all those elements of the universal set whose membership value in fuzzy set A are greater than or equal to the specified value of α. The set is,

$$A^\alpha = \{x \in X | \mu_A \geq \alpha\} \qquad (7.26)$$

For instance, let the universe of discourse $X = \{x_0, x_1, x_2, \ldots, x_{10}\}$, and α-cut of fuzzy set A and its membership function $\mu_A(x)$ is shown in Fig. (7.1). Then

$$A^\alpha = \{x_1, x_2, \ldots, x_9\} \qquad (7.27)$$

Note, if for any fuzzy set A if $\alpha_1 < \alpha_2$, then it implies $A_2^\alpha \subseteq A_1^\alpha$.

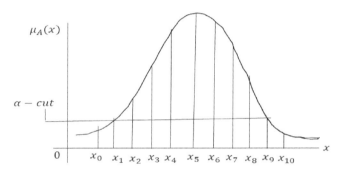

Figure 7.1: α-cut Membership Function.

7.2.3.4 Strong α-cut

It means the fuzzy set A is the crisp set that contains all those elements of the universal set whose membership value in fuzzy set A is greater than α for $\alpha \in [0,1]$. That is,

$$A^{\alpha^+} = \{x \in X | \mu_A > \alpha\} \qquad (7.28)$$

In the above discussed example, $A^{\alpha^+} = \{x_2, x_3, \ldots, x_8\}$

7.2.3.5 Level Set

The set of all levels $\alpha \in [0,1]$ that represent distinct α-cuts of a given fuzzy set A is called a level set of A and it is denoted by $\bigwedge(A)$. That is,

$$\bigwedge(A) = \{\alpha : \mu_A(x) = \alpha \text{ for some } x \in X\} \qquad (7.29)$$

Example 106 *Let $X = \{1, 2, \ldots, 5\}$ and fuzzy set $A = \{(1, 0.7), (2, 0.3), (3, 0.5), (4, 0.9), (5, 0.6)\}$. Now we will calculate strong α-cut for $\alpha = 0.3, 0.5, \& 0.9$ and also the level set of A. As strong α-cut set of A is*

$$A^{\alpha^+} = \{x \in X | \mu_A > \alpha\}.$$

Therefore,
$$A^{0.3^+} = \{x \in X | \mu_A > 0.3\} = \{1,3,4,5\},$$
and
$$A^{0.5^+} = \{x \in X | \mu_A > 0.5\} = \{1,4,5\}, \ A^{0.9^+} = \{x \in X | \mu_A > 0.9\} = \Phi$$

Finally, the level set of A is
$$\bigwedge(A) = \{\alpha : \mu_A(x) = \alpha \text{ for some } x \in X\} = \{0.3, 0.5, 0.6, 0.7, 0.9\}.$$

Example 107 Consider two fuzzy sets A and B on universal set $X = \{1,2,\ldots,6\}$, where $A = \left\{\dfrac{0.3}{1} \oplus \dfrac{1}{2} \oplus \dfrac{0.5}{3} \oplus \dfrac{0.6}{4} \oplus \dfrac{0.2}{5} \oplus \dfrac{0.6}{6}\right\}$,
$B = \left\{\dfrac{0.9}{1} \oplus \dfrac{0.5}{2} \oplus \dfrac{0.8}{3} \oplus \dfrac{0.4}{4} \oplus \dfrac{0.7}{5} \oplus \dfrac{0.3}{6}\right\}.$

We will find $(A \bigcup B)^{0.7}$ and $(A \bigcap B)^{0.3}$. According to definition,
$\mu_{A \bigcup B}(x) = \max\{\mu_A(x), \mu_B(x)\}, \forall x \in X.$
$\mu_{A \bigcup B}(1) = \max\{\mu_A(1), \mu_B(1)\} = \max\{0.3, 0.9\} = 0.9.$
$\mu_{A \bigcup B}(2) = \max\{\mu_A(2), \mu_B(2)\} = \max\{1, 0.5\} = 1,$ and so on. Finally, we have
$$\mu_{A \bigcup B} = \left\{\dfrac{0.9}{1} \oplus \dfrac{1.0}{2} \oplus \dfrac{0.8}{3} \oplus \dfrac{0.6}{4} \oplus \dfrac{0.7}{5} \oplus \dfrac{0.6}{6}\right\}.$$

By definition of α-cut set $A^\alpha = \{x \in X | \mu_A \geq \alpha\}$. Hence $(A \bigcup B)^{0.7} = \{1,2,3,5\}$.

Recall, the definition $\mu_{A \bigcap B}(x) = \min\{\mu_A(x), \mu_B(x)\}, \forall x \in X.$ Using this defination, we have
$$\mu_{A \bigcap B} = \left\{\dfrac{0.3}{1} \oplus \dfrac{0.5}{2} \oplus \dfrac{0.5}{3} \oplus \dfrac{0.4}{4} \oplus \dfrac{0.2}{5} \oplus \dfrac{0.3}{6}\right\}.$$ Hence $(A \bigcap B)^{0.3} = \{1,2,3,4,6\}$.

Example 108 Let fuzzy sets A and B on universal set $X = \{20, 40, \ldots, 100\}$, where $A = \left\{\dfrac{0.5}{20} \oplus \dfrac{0.65}{40} \oplus \dfrac{0.85}{60} \oplus \dfrac{1.0}{80} \oplus \dfrac{1.0}{100}\right\}$,
$B = \left\{\dfrac{0.35}{20} \oplus \dfrac{0.5}{40} \oplus \dfrac{0.75}{60} \oplus \dfrac{0.90}{80} \oplus \dfrac{1.0}{100}\right\}.$

We will find $(\overline{A \bigcup B})^{0.15}$ and $(A \bigcup B)^{0.5^+}$. Recall the definition of membership function of standard compliment of A is $\mu_{\bar{A}} = 1 - \mu_A(x), \forall x \in X$. Therefore,
$$\bar{A} = \left\{\dfrac{0.5}{20} \oplus \dfrac{0.35}{40} \oplus \dfrac{0.15}{60} \oplus \dfrac{0.0}{80} \oplus \dfrac{0.0}{100}\right\},$$
$$\bar{B} = \left\{\dfrac{0.65}{20} \oplus \dfrac{0.50}{40} \oplus \dfrac{0.25}{60} \oplus \dfrac{0.1}{80} \oplus \dfrac{0.0}{100}\right\}.$$

$$\overline{(A \bigcup B)} = \overline{A} \bigcap \overline{B} = \min[\mu_A(x), \mu_B(x)]$$

7.3 Fuzzy Numbers

There are various types of fuzzy numbers available in the literature. In this section, two most popular and commonly used fuzzy numbers are defined.

7.3.1 Triangular Fuzzy Number

Using a triplet say (a,b,c) a triangular fuzzy number $\tilde{x}$ can be defined whose membership function $\omega_{\tilde{x}}(y)$ is (see Fig. (7.2)):

$$\omega_{\tilde{x}}(y) = \begin{cases} 0, & \text{if } y \leq a \\ \frac{y-a}{b-a}, & \text{if } a \leq y \leq b \\ \frac{c-y}{c-b}, & \text{if } b \leq y \leq c \\ 0, & \text{if } y \geq c \end{cases}$$

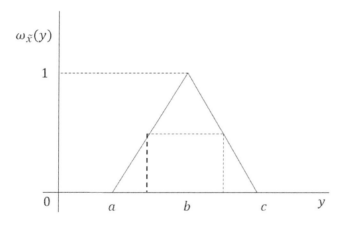

Figure 7.2: Triangular Fuzzy Number.

7.3.2 Trapezoidal Fuzzy Number

A trapezoidal fuzzy number $\tilde{x}$ is denoted by (a,b,c,d) whose membership function $\omega_{\tilde{x}}(y)$ is (see Fig. (7.3)):

$$\omega_{\tilde{x}}(y) = \begin{cases} 0, & \text{if } y \leq a \\ \frac{y-a}{b-a}, & \text{if } a \leq y \leq b \\ 1, & \text{if } b \leq y \leq c \\ \frac{d-y}{d-c}, & \text{if } c \leq y \leq d \\ 0, & \text{if } y \geq d \end{cases}$$

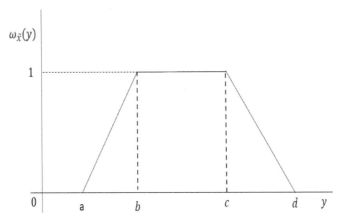

Figure 7.3: Trapezoidal Fuzzy Number.

7.4 Defuzzification Methods

The process of converting fuzzy quantity into crisp quantity is called as "Defuzzification". It selects the best crisp value out of the fuzzy set. There are several methods of defuzzification such as α-cut, ranking method, the center of sums (COS), the center of gravity (COG), mean of maximum (MOM), and weighted average method. Some of these methods are discussed in the following sections.

7.4.1 Center of Sums Method

Most commonly used defuzzification method is center of sums (COS) method. This method counts the overlapping area twice. Let y be any fuzzy number, then the defuzzified value of y is defined as:

$$y^* = \frac{\sum_{i=1}^{N} y_i \sum_{k=1}^{n} \omega_{A_k}(y_i)}{\sum_{i=1}^{N} \sum_{k=1}^{n} \omega_{A_k}(y_i)}$$

where, n denotes number of fuzzy sets, N denotes number of fuzzy variables, and $\omega_{A_k}(y_i)$ denotes the membership function for the kth fuzzy set.

Example 109 *Let a triangular fuzzy number be $\tilde{y} = (3,5,7)$. Find the defuzzified value using center of sums method.*

Let the area of the fuzzy number $\tilde{y}$ be (see Fig. (7.4)):

$$A = \frac{1}{2} \times (7-3) \times 1 = \frac{4}{2} = 2$$

and the center of the fuzzy set is say $\bar{x} = \frac{7+3}{2} = 5$. Now, the defuzzified value will be:

$$y^* = \frac{A\bar{x}}{A} = \frac{2 \times 5}{2} = 5$$

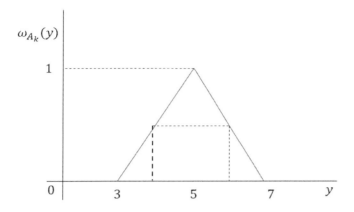

Figure 7.4: Fuzzy Set.

Example 110 *Let there be two trapezoidal fuzzy sets say $\tilde{y}_1 = (1,3,7,8)$ and $\tilde{y}_2 = (3,4,8,9)$. Find the defuzzified value using center of sums method.*

Let the area of the fuzzy sets be (see Fig. (7.5)):

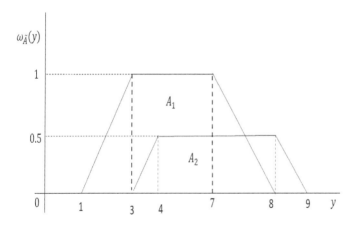

Figure 7.5: Fuzzy Sets.

$$A_1 = \frac{1}{2}[(8-1)+(7-3)] \times 1 = 5.5$$

$$A_2 = \frac{1}{2}[(9-3)+(8-4)] \times 0.5 = 2.5$$

and the centers of the fuzzy sets are say

$$\bar{x}_1 = \frac{7+3}{2} = 5; \quad \bar{x}_2 = \frac{8+4}{2} = 6$$

Now, the defuzzified value will be:

$$y^* = \frac{A_1 \bar{x}_1 + A_2 \bar{x}_2}{A_1 + A_2} = \frac{5.5 \times 5 + 2.5 \times 6}{11} = 3.43$$

7.4.2 Center of Gravity Method

Based on the center of gravity (COG) of fuzzy set this method provides a crisp value. The total area of the membership function distribution used to represent the combined control action is divided into a number of sub-areas. The area and the center of gravity or centroid of each sub-area is calculated and then the summation of all these sub-areas is taken to find the defuzzified value for a discrete fuzzy set.

The defuzzified value denoted by y^* (say) for discrete membership function using COG is defined as:

$$y^* = \frac{\sum_{i=1}^{n} y_i \omega(y_i)}{\sum_{i=1}^{n} \omega(y_i)}$$

where y_i denotes the sample element, $\omega(y_i)$ is the membership function, and n represents the number of elements in the sample.

The defuzzified value y^* for continuous membership function is defined as:

$$y^* = \frac{\int y \omega_A(y) dy}{\int \omega_A(y) dy}$$

Example 111 *Let the fuzzy trapezoidal set be $\tilde{y} = (3, 4, 7, 9)$. Find the defuzzified value using center of gravity method.*

The defuzzified value y^* using center of gravity is defined as:

$$y^* = \frac{\sum_{i=1}^{N} A_i \bar{x}_i}{\sum_{i=1}^{N} A_i}$$

where $N \to$ denotes the number of sub-areas
$A_i \to$ area of ith sub-area
$\bar{x}_i \to$ centroid area of ith sub-area

The total area of the aggregated fuzzy set (see Fig. (7.6)) is divided into three sub-areas. Now, to convert fuzzy set into crisp using COG, we have to compute area and centroid area of each sub-area. Calculation is shown below:

Total area of sub-area 'a' $= \frac{1}{2} \times 1 \times 1 = 0.5$
Total area of sub-area 'b' $= (7-4) \times 1 = 3$
Total area of sub-area 'c' $= \frac{1}{2} \times 2 \times 1 = 1$

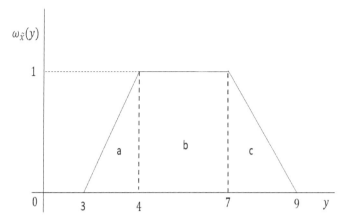

Figure 7.6: Fuzzy Set.

Centroid of sub-area 'a'=$\frac{(3+4+4)}{3}=3.67$
Centroid of sub-area 'b'=$\frac{(7+4)}{2}=5.5$
Centroid of sub-area 'c'=$\frac{(7+7+9)}{3}=7.67$

Now, the defuzzified set will be:

$$y^* = \frac{(0.5 \times 3.67) + (3 \times 5.5) + (1 \times 7.67)}{(0.5 + 3 + 1)} = 5.78$$

Similarly, the two fuzzy sets can be converted into crisp value using COG method.

7.4.3 α-cut Method

Let $\tilde{x} = (a,b,c)$ a fuzzy number. An α-cut for $\tilde{x}$, $\tilde{x}_\alpha$ is computed as:

$$\alpha = \frac{x-a}{b-a} \Rightarrow \tilde{x}_\alpha^L = x = (b-a)\alpha + a$$

$$\alpha = \frac{c-x}{c-b} \Rightarrow \tilde{x}_\alpha^U = x = c - (c-b)\alpha$$

where $\tilde{x}_\alpha = [\tilde{x}_\alpha^L, \tilde{x}_\alpha^U]$ is the corresponding α-cut and $\tilde{x}_\alpha^L$, $\tilde{x}_\alpha^U$ represents the lower and the upper bounds respectively (see Fig. (7.7)).

For prescribed value of α, the method can be used to convert the fuzzy numbers into crisp numbers.

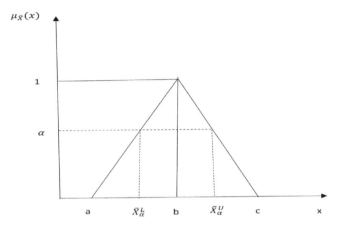

Figure 7.7: Graphical Representation of α-cut.

Example 112 *Let the fuzzy set be $\tilde{y} = (4, 6, 8)$. Find the crisp value using α-cut method.*

To convert the fuzzy set into crisp set using α-cut method for minimization type objective function and $\leq$ type constraints, use lower bound α-cut formula, i.e.,

$$y^* = (6-4)\alpha + 4 = 2\alpha + 4$$

And for the maximization type objective function and $\geq$ type constraints, use upper bound α-cut formula, i.e.,

$$y^* = 8 - (8-6)\alpha = 8 - 2\alpha$$

7.5 Solving Optimization Problem with Fuzzy Numbers using LINGO-18

In this section, three cases of the linear optimization problem are solved through LINGO-18 software. The first case is when objective function coefficients are fuzzy numbers, second when constraint coefficient parameters, and third when RHS parameters are represented as fuzzy numbers. The conversion of fuzzy numbers into crisp is shown by all the three defuzzification methods discussed in Section (7.4).

7.5.1 Case I: When Coefficients in Objective Function are Fuzzy Numbers

Consider an uncertain LPP:

$$\left.\begin{array}{l}\max\ Z = (1,2,4)y_1 + (1,3,5)y_2 \\ \text{subject to } y_1 + y_2 \leq 1 \\ \qquad 3y_1 + y_2 \leq 4 \\ \qquad y_1, y_2 \geq 0\end{array}\right\} \qquad (7.30)$$

(i) Solution by converting the fuzzy numbers into crisp using COS method.
Defuzzified value of $(1,2,4) = \frac{(1/2 \times 3 \times 1)5/2}{3/2} = 2.5$
Defuzzified value of $(1,3,5) = \frac{(1/2 \times 4 \times 1)6/2}{2} = 3$
Now, the LINGO-18 codes for the crisp LPP will be:

```
MODEL:
MAX= 2.5 * Y1 + 3 * Y2;
Y1 + Y2 <= 1;
3 * Y1 + Y2 <= 4;
END
```

The global optimal solution is $y_1 = 0, y_2 = 1$ with maximum $Z = 3$.

(ii) Solution by converting the fuzzy numbers into crisp using COG method.
Defuzzified value of $(1,2,4) = \frac{(1/2 \times 3 \times 1)(1+2+4)/3}{3/2} = 2.33$
Defuzzified value of $(1,3,5) = \frac{(1/2 \times 4 \times 1)(1+3+5)/3}{2} = 3$
Now, the LINGO-18 codes for the crisp LPP will be:

```
MODEL:
MAX= 2.33 * Y1 + 3 * Y2;
Y1 + Y2 <= 1;
3 * Y1 + Y2 <= 4;
END
```

The global optimal solution is $y_1 = 0, y_2 = 1$ with maximum $Z = 3$.

(iii) Solution by converting the fuzzy numbers into crisp using α-cut method.
Defuzzified value of $(1,2,4) = 4 - 2\alpha$
Defuzzified value of $(1,3,5) = 5 - 2\alpha$
Now, the LINGO-18 codes for the crisp LPP will be:

```
MODEL:
MAX= (4 - 2 * A) * Y1 + (5 - 2 * A) * Y2;
Y1 + Y2 <= 1;
3 * Y1 + Y2 <= 4;
A = 0.5
END
```

The global optimal solution at $\alpha = 0.5$ is $y_1 = 0, y_2 = 1$ with maximum $Z = 4$.

7.5.2 Case II: When Constraint Coefficients are Fuzzy Numbers

Consider an uncertain LPP:

$$\left.\begin{aligned}\max\ Z &= 2y_1 + 3y_2 \\ \text{subject to}\ (1,2,3)y_1 + (2,4,6)y_2 &\leq 1 \\ (2,4,5)y_1 + (1,3,5)y_2 &\leq 4 \\ y_1, y_2 &\geq 0\end{aligned}\right\} \quad (7.31)$$

(i) Solution by converting the fuzzy numbers into crisp using COS method.
   Defuzzified value of $(1,2,3) = 2$
   Defuzzified value of $(2,4,6) = 4$
   Defuzzified value of $(2,4,5) = 3.5$
   Defuzzified value of $(1,3,5) = 3$
   Now, the LINGO-18 codes for the crisp LPP will be:

```
MODEL:
MAX= 2 * Y1 + 3 * Y2;
2 * Y1 + 4 * Y2 <= 1;
3.5 * Y1 + 3 * Y2 <= 4;
END
```

The global optimal solution is $y_1 = 0.5, y_2 = 0$ with maximum $Z = 1$.

(ii) Solution by converting the fuzzy numbers into crisp using COG method.
   Defuzzified value of $(1,2,3) = 2$
   Defuzzified value of $(2,4,6) = 4$
   Defuzzified value of $(2,4,5) = 3.67$
   Defuzzified value of $(1,3,5) = 3$
   Now, the LINGO-18 codes for the crisp LPP will be:

```
MODEL:
MAX= 2 * Y1 + 3 * Y2;
2 * Y1 + 4 * Y2 <= 1;
3.67 * Y1 + 3 * Y2 <= 4;
END
```

The global optimal solution is $y_1 = 0.5, y_2 = 0$ with maximum $Z = 1$.

(iii) Solution by converting the fuzzy numbers into crisp using α-cut method.
Defuzzified value of $(1,2,3) = \alpha + 1$
Defuzzified value of $(2,4,6) = 2\alpha + 2$
Defuzzified value of $(2,4,5) = 2\alpha + 2$
Defuzzified value of $(1,3,5) = 2\alpha + 1$
Now, the LINGO-18 codes for the crisp LPP will be:

```
MODEL:
MAX= 2 * Y1 + 3 * Y2;
(A + 1) * Y1 + (2 * A + 2) * Y2 <= 1;
(2 * A + 2) * Y1 + (2 * A + 1) * Y2 <= 4;
A = 0.5
END
```

The global optimal solution at $\alpha = 0.5$ is $y_1 = 0.67, y_2 = 0$ with maximum $Z = 1.33$.

7.5.3 Case III: When RHS Parameters are Fuzzy Numbers

Consider an uncertain LPP:

$$\left. \begin{array}{l} \max \ Z = 2y_1 + 3y_2 \\ \text{subject to } y_1 + y_2 \leq (1,3,4) \\ \quad\quad\quad\quad 3y_1 + y_2 \leq (2,4,8) \\ \quad\quad\quad\quad y_1, y_2 \geq 0 \end{array} \right\} \quad (7.32)$$

(i) Solution by converting the fuzzy numbers into crisp using COS method.
Defuzzified value of $(1,3,4) = 2.5$
Defuzzified value of $(2,4,8) = 5$
Now, the LINGO-18 codes for the crisp LPP will be:

```
MODEL:
MAX= 2 * Y1 + 3 * Y2;
Y1 + Y2 <= 2.5;
3 * Y1 + Y2 <= 5;
END
```

The global optimal solution is $y_1 = 0, y_2 = 2.5$ with maximum $Z = 7.5$.

(ii) Solution by converting the fuzzy numbers into crisp using COG method.
Defuzzified value of $(1,3,4) = 2.67$
Defuzzified value of $(2,4,8) = 4.67$
Now, the LINGO-18 codes for the crisp LPP will be:

```
MODEL:
MAX= 2 * Y1 + 3 * Y2;
Y1 + Y2 <= 2.67;
3 * Y1 + Y2 <= 4.67;
END
```

The global optimal solution is $y_1 = 0, y_2 = 2.67$ with maximum $Z = 8.01$.

(iii) Solution by converting the fuzzy numbers into crisp using α-cut method.
Defuzzified value of $(1,3,4) = 2\alpha + 1$
Defuzzified value of $(2,4,8) = 2\alpha + 2$
Now, the LINGO-18 codes for the crisp LPP will be:

```
MODEL:
MAX= 2 * Y1 + 3 * Y2;
Y1 + Y2 <= (2 * A + 1);
3 * Y1 + Y2 <= (2 * A + 2);
A = 0.5
END
```

The global optimal solution at $\alpha = 0.5$ is $y_1 = 0, y_2 = 2$ with maximum $Z = 6$.

7.6 Stochastic Optimization Problem

Consider an LPP in the following form:

$$\left. \begin{array}{l} \min \ f(\underline{x}) = \sum_{j=1}^{n} c_j x_j \\ \text{subject to} \ \sum_{j=1}^{n} a_{ij} x_j \leq b_i, \ i = 1,2,\ldots,m \\ x_j \geq 0, \ j = 1,2,\ldots,n \end{array} \right\} \quad (7.33)$$

In standard LPP the parameters $\underline{C}, A$ and $\underline{b}$ are assumed to be deterministic and known. If some or all of these parameters are random variables then their values are not absolute, and uncertainty prevails. The study of such linear optimization problems under the above situation is known as "Optimization under Uncertainty", or "Stochastic Programming." In the following section various cases of stochastic programming are discussed.

7.6.1 Situation I: Parameters in Objective Function C_j are Random Variables

For simplicity, we assume that C_j are random variables having a normal distribution with mean $\bar{C}_j$, variance $V(C_j)$ and variance-covariance matrix Σ as:

$$\Sigma = \begin{pmatrix} V(C_1) & Cov(C_1, C_2) & \cdots & Cov(C_1, C_n) \\ Cov(C_2, C_1) & V(C_2) & \cdots & Cov(C_2, C_n) \\ \vdots & \vdots & \ddots & \vdots \\ Cov(C_n, C_1) & Cov(C_n, C_1) & \cdots & V(C_n) \end{pmatrix}$$

The objective function $f(\underline{x}) = \sum_{j=1}^{n} c_j x_j$ being a linear combination of normally distributed random variables will also be a random variable normally distributed with mean

$$\bar{f}(\underline{x}) = \sum_{j=1}^{n} \bar{C}_j x_j \text{ and } V(f(\underline{x})) = \underline{x}' \Sigma \underline{x}$$

Thus, a deterministic equivalent of the objective function $f(\underline{x})$ may be given as

$$f(\underline{x}) = k_1 \sum_{j=1}^{n} \bar{C}_j x_j + k_2 \sqrt{\underline{x}' \Sigma \underline{x}}$$

where k_1 & $k_2 \geq 0$ are weights assigned to $\bar{f}$ and $S.D.(f)$ accordingly to their relative importance.

If a_{ij} and b_j are known constants, there will be no uncertainty in the constraints, and hence the deterministic equivalent of the stochastic LPP with only C_j as normally distributed random variables given as:

$$\left.\begin{array}{l} \min \ f(\underline{x}) = k_1 \sum_{j=1}^{n} \bar{C}_j x_j + k_2 \sqrt{\underline{x}' \Sigma \underline{x}} \\ \text{subject to} \ \sum_{j=1}^{n} a_{ij} x_j \leq b_i, \ i = 1, 2, \ldots, m \\ x_j \geq 0, \ j = 1, 2, \ldots, n \end{array}\right\} \quad (7.34)$$

In case C_j are independently distributed, the $Cov(C_i, C_j) = 0$; $i \neq j = 1, 2, \ldots, n$ and

$$\underline{x}' \Sigma \underline{x} = (x_1, x_2, \ldots, x_n) \begin{pmatrix} V(C_1) & 0 & \cdots & 0 \\ 0 & V(C_2) & \cdots & 0 \\ \vdots & \vdots & \ddots & \vdots \\ 0 & 0 & \cdots & V(C_n) \end{pmatrix} \begin{pmatrix} x_1 \\ x_2 \\ \vdots \\ x_n \end{pmatrix} = \sum_{j=1}^{n} V(C_j) x_j^2$$

Moreover, the deterministic equivalence form will become

$$\left.\begin{array}{l} \min \ f(\underline{x}) = k_1 \sum_{j=1}^{n} \bar{C}_j x_j + k_2 \sqrt{\sum_{j=1}^{n} V(C_j) x_j^2} \\ \text{subject to} \ \sum_{j=1}^{n} a_{ij} x_j \leq b_i, \ i = 1, 2, \ldots, m \\ x_j \geq 0, \ j = 1, 2, \ldots, n \end{array}\right\} \quad (7.35)$$

Similarly, the deterministic equivalence form for the maximization case can be given as:

$$\left.\begin{array}{l} \max \ f(\underline{x}) = k_1 \sum_{j=1}^{n} \bar{C}_j x_j - k_2 \sqrt{\sum_{j=1}^{n} V(C_j) x_j^2} \\ \text{subject to} \ \sum_{j=1}^{n} a_{ij} x_j \leq b_i, \ i = 1, 2, \ldots, m \\ x_j \geq 0, \ j = 1, 2, \ldots, n \end{array}\right\} \quad (7.36)$$

7.6.2 Situation II: Availability/Requirement Vector b_i are Probabilistic

When only b_i are random variables the LP is called Chance Constrained LPP. Charnes & Cooper [8]; [9] proposed chance constrained programming which considers random data variations and allows constraints violations up to specified probability limits. The probabilistic chance constraints are

$$P\left(\sum_{j=1}^{n} a_{ij}x_j \leq b_i\right) \geq \beta_i = 1 - F_i\left(\sum_{j=1}^{n} a_{ij}x_j\right), \quad i = 1,2,\ldots,m$$

$$\text{or } F_i\left(\sum_{j=1}^{n} a_{ij}x_j\right) \leq 1 - \beta_i, \quad i = 1,2,\ldots,m \quad (7.37)$$

where $F(\tau)$ denotes the distribution function of b_i and $\beta_1, \beta_2, \ldots, \beta_m$ are given probabilities of the extents to which constraint violations are admitted. The inequalities (7.37) are called chance constraints which means that the ith constraint may be violated, but at most $(1-\beta_i)$ proportion.

Let $Z_{1-\beta_i}$ denote the maximum of τ such that $\tau = F_i^{-1}(1-\beta_i)$, and then the inequality (7.37) can be expressed as

$$\sum_{j=1}^{n} a_{ij}x_j \leq Z_{1-\beta_i}, \quad i = 1,2,\ldots,m$$

In particular, if b_i, $i = 1,2,\ldots,m$ are i.i.d, random variables have a normal distribution with mean $\bar{b}_i$ and variance $\sigma_{b_i}^2$. It follows that

$$P\left(\frac{\sum_{j=1}^{n} a_{ij}x_j - \bar{b}_i}{\sigma_{b_i}} \leq \frac{b_i - \bar{b}_i}{\sigma_{b_i}}\right) \geq \beta_i, \quad i = 1,2,\ldots,m$$

$$\Rightarrow P\left(\frac{b_i - \bar{b}_i}{\sigma_{b_i}} \geq \frac{\sum_{j=1}^{n} a_{ij}x_j - \bar{b}_i}{\sigma_{b_i}}\right) \geq \beta_i, \text{ or } 1 - \Phi\frac{\sum_{j=1}^{n} a_{ij}x_j - \bar{b}_i}{\sigma_{b_i}}$$

$$\Rightarrow P\left(Z \geq \frac{\sum_{j=1}^{n} a_{ij}x_j - \bar{b}_i}{\sigma_{b_i}}\right) \geq \beta_i, \quad i = 1,2,\ldots,m \quad (7.38)$$

where $Z \sim N(0,1)$. If Z_i is the value of $Z \sim N(0,1)$ such that $P[Z \geq Z_i] = \beta_i$ or $P[Z \leq Z_i] = 1 - \beta_i$, then Eqn. (7.38) will be true only if

$$\frac{\sum_{j=1}^{n} a_{ij}x_j - \bar{b}_i}{\sigma_{b_i}} \leq Z_i$$

$$\sum_{j=1}^{n} a_{ij}x_j - \bar{b}_i \leq Z_i\sigma_{b_i}$$

$$\Rightarrow \sum_{j=1}^{n} a_{ij}x_j \leq \bar{b}_i + Z_i\sigma_{b_i}, \quad \forall\, i$$

or $\sum_{j=1}^{n} a_{ij}x_j \leq \bar{b}_i + \sigma_{b_i}\Phi^{-1}(1-\beta_i), \ \forall \ i$

where Φ is the distribution function of the standard normal.
Thus, the problem may be expressed as the following deterministic LPP

$$\left.\begin{aligned} \min \ f(\underline{x}) &= k_1 \sum_{j=1}^{n} C_j x_j \\ \text{subject to} \ & \sum_{j=1}^{n} a_{ij}x_j \leq \bar{b}_i + Z_i \sigma_{b_i}, \ i=1,2,\ldots,m \\ \text{or} \ & \sum_{j=1}^{n} a_{ij}x_j \leq \bar{b}_i + \sigma_{b_i}\Phi^{-1}(1-\beta_i), \ \forall \ i \\ & x_j \geq 0, \ j=1,2,\ldots,n \end{aligned}\right\} \quad (7.39)$$

7.6.3 Situation III: When a_{ij} are Random Variables

Let a_{ij}, $i = 1, 2, \ldots, m$; $j = 1, 2, \ldots, n$ are normally distributed with mean $\bar{a}_{ij}$ and variance $\sigma^2_{a_{ij}}$. Also, let $Cov(a_{ij}, a_{kl})$ denote the covariance between a_{ij} and a_{kl}. Define $d_i = \sum_{j=1}^{n} a_{ij} x_j$ being a linear combination of normal random variables d_i; $i = 1, 2, \ldots, m$ are also normally distributed with mean $\bar{d}_i = \sum \bar{a}_{ij} x_j$, $\forall \ i$ and variance $\sigma^2_{d_i} = \underline{x}' \sum_i \underline{x}$, $\forall \ i$, where Σ_i is the variance-covariance matrix of a_{ij} given as:

$$\Sigma_i = \begin{pmatrix} \sigma^2_{a_{i1}} & Cov(a_{i1}, a_{i2}) & \cdots & Cov(a_{i1}, a_{in}) \\ Cov(a_{i2}, a_{i1}) & \sigma^2_{a_{i2}} & \cdots & Cov(a_{i2}, a_{in}) \\ \vdots & \vdots & \ddots & \vdots \\ Cov(a_{in}, a_{i1}) & Cov(a_{in}, a_{i2}) & \cdots & \sigma^2_{a_{in}} \end{pmatrix}$$

The probabilistic constraints $P\left(\sum_{j=1}^{n} a_{ij}x_j \leq \bar{b}_i\right) \geq \beta_i$, $i = 1, 2, \ldots, m$ can now be expressed as

$$P(d_i \leq b_i) \geq \beta_i, \ i = 1, 2, \ldots, m$$

$$\Rightarrow P\left(\frac{d_i - \bar{d}_i}{\sigma_{d_i}} \leq \frac{b_i - \bar{d}_i}{\sigma_{d_i}}\right) \geq \beta_i, \ i = 1, 2, \ldots, m$$

$$\Rightarrow P\left(Z \leq \frac{b_i - \bar{d}_i}{\sigma_{d_i}}\right) \geq \beta_i, \ i = 1, 2, \ldots, m \quad (7.40)$$

where $Z \sim N(0, 1)$. If Z_i is the value of $Z \sim N(0, 1)$ such that $P[Z \leq Z_i] = \beta_i$, then Eqn. (7.40) will be true only if

$$\frac{b_i - \bar{d}_i}{\sigma_{d_i}} \geq Z_i$$

$$b_i - \bar{d}_i \geq Z_i \sigma_{d_i} \Rightarrow \bar{d}_i + Z_i \sigma_{d_i} \leq b_i, \ \forall \ i$$

This gives the deterministic equivalent of the chance constraints LPP as:

$$\left. \begin{array}{l} \min \ f(\underline{x}) = k_1 \sum_{j=1}^{n} C_j x_j \\ \text{subject to} \ \sum_{j=1}^{n} \bar{a}_{ij} x_j + Z_i \sqrt{\underline{x}' \Sigma_i \underline{x}} \leq \bar{b}_i, \ i = 1, 2, \ldots, m \\ x_j \geq 0, \ j = 1, 2, \ldots, n \end{array} \right\} \quad (7.41)$$

In case a_{ij} are independently distributed the covariance terms in Σ_i are zero and $\underline{x}' \Sigma_i \underline{x} = \sum_{j=1}^{n} \sigma_{a_{ij}}^2 x_j^2$. This gives deterministic constraints as:

$$\sum_{j=1}^{n} \bar{a}_{ij} x_j + Z_i \sqrt{\sum_{j=1}^{n} \sigma_{a_{ij}}^2 x_j^2} \leq \bar{b}_i, \ i = 1, 2, \ldots, m$$

Moreover, we have the following deterministic NLPP to solve

$$\left. \begin{array}{l} \min \ f(\underline{x}) = k_1 \sum_{j=1}^{n} C_j x_j \\ \text{subject to} \ \sum_{j=1}^{n} \bar{a}_{ij} x_j + Z_i \sqrt{\sum_{j=1}^{n} \sigma_{a_{ij}}^2 x_j^2} \leq \bar{b}_i, \ i = 1, 2, \ldots, m \\ x_j \geq 0, \ j = 1, 2, \ldots, n \end{array} \right\} \quad (7.42)$$

7.6.4 Situation IV: General Case—All Parameters are Random Variables

Consider the chance-constrained LPP:

$$\left. \begin{array}{l} \min \ f(\underline{x}) = \sum_{j=1}^{n} C_j x_j \\ \text{subject to} \ P\left(\sum_{j=1}^{n} a_{ij} x_j \leq b_i\right) \geq \beta_i, \ i = 1, 2, \ldots, m \\ x_j \geq 0, \ j = 1, 2, \ldots, n \end{array} \right\} \quad (7.43)$$

where C_j, a_{ij} and b_i are random variables. For simplicity, we assume that $C_j \sim N(\bar{C}_j, \sigma_{C_j}^2); \ a_{ij} \sim N(\bar{a}_{ij}, \sigma_{a_{ij}}^2)$ and $b_i \sim N(\bar{b}_i, \sigma_{b_i}^2)$.

As random variables C_j appear only in the objective function the deterministic equivalent of the objective function, $f(\underline{x})$ can be obtained as in the case I. We have to minimise

$$f(\underline{x}) = k_1 \sum_{j=1}^{n} \bar{C}_j x_j + k_2 [\underline{x}' \Sigma \underline{x}]^{1/2} \tag{7.44}$$

where the symbols have their usual meaning.

The probabilistic constraints can be expressed as: $P(h_i \leq 0) \geq \beta_i$, $i = 1, 2, \ldots, m$.
where $h_i = \sum_{j=1}^{n} a_{ij} x_j - b_i$, $i = 1, 2, \ldots, m$ are normal distributed random variables with mean $\bar{h}_i$ and variance $\sigma_{h_i}^2$.

Now, let

$$q_{ik} = a_{ik}; \ k = 1, 2, \ldots, n; \quad q_{ik} = b_i; \ k = n+1$$
$$\text{and } y_k = x_k; \ k = 1, 2, \ldots, n; \quad y_k = -1; \ k = n+1$$

Then,

$$h_i = \sum_{j=1}^{n} a_{ij} x_j - b_i = \sum_{k=1}^{n} q_{ik} y_k, \ i = 1, 2, \ldots, m$$

$$\bar{h}_i = \sum_{k=1}^{n} \bar{q}_{ik} y_k = \sum_{j=1}^{n} \bar{a}_{ij} x_j - b_i =, \ \forall \, i$$

Also,

$$\sigma_{hi}^2 = \underline{y}' \Sigma_i \underline{y}, \text{ where } \underline{y}' = (y_1, y_2, \ldots, y_{n+1}) = (x_1, x_2, \ldots, x_{n+1}, -1)$$

and

$$\Sigma_i = \begin{pmatrix} V(q_{i1}) & Cov(q_{i1}, q_{i2}) & \cdots & Cov(q_{i1}, q_{in+1}) \\ Cov(q_{i2}, q_{i1}) & V(q_{i2}) & \cdots & Cov(q_{i2}, q_{in+1}) \\ \vdots & \vdots & \ddots & \vdots \\ Cov(q_{in+1}, q_{i1}) & Cov(q_{in+1}, q_{i2}) & \cdots & V(q_{in+1}) \end{pmatrix}$$

The probabilistic constraints $P(h_i \leq 0) \geq p_i$, $i = 1, 2, \ldots, m$ can now be expressed as:

$$\Rightarrow P\left(\frac{h_i - \bar{h}_i}{\sigma_{h_i}} \leq -\frac{\bar{h}_i}{\sigma_{h_i}}\right) \geq \beta_i, \ i = 1, 2, \ldots, m$$

$$\Rightarrow P\left(Z \leq -\frac{\bar{h}_i}{\sigma_{h_i}}\right) \geq \beta_i, \ i = 1, 2, \ldots, m \tag{7.45}$$

where $Z \sim N(0,1)$. If Z_i is the value of $Z \sim N(0,1)$ such that $P[Z \leq Z_i] = \beta_i$, then Eqn. (7.45) will be true only if

$$-\frac{\bar{h}_i}{\sigma_{h_i}} \geq Z_i$$

$$-\bar{h}_i \geq Z_i \sigma_{h_i} \Rightarrow \bar{h}_i + Z_i \sigma_{h_i} \leq 0, \ \forall \ i \tag{7.46}$$

Equations (7.44) and (7.46) give the equivalent deterministic of the given chance-constrained LPP as:

$$\left.\begin{array}{l} \min \ f(\underline{x}) = k_1 \sum_{j=1}^{n} \bar{C}_j x_j + k_2 \sqrt{\underline{x}' \Sigma \underline{x}} \\ \text{subject to } \bar{h}_i + Z_i \sigma_{h_i} \leq 0, \ \forall \ i \\ x_j \geq 0, \ j = 1, 2, \ldots, n \end{array}\right\} \tag{7.47}$$

where the symbols have their usual meanings.

7.7 Some Numerical Examples using LINGO-18

Example 113 *Consider the LPP with uncertain coefficients in objective function.*

$$\left.\begin{array}{l} \max \ f(\underline{x}) = \bar{40}x_1 + \bar{15}x_2 \\ \text{subject to } 6x_1 + 5x_2 \leq 420 \\ \quad 4x_1 + 9x_2 \leq 410 \\ \quad 2x_1 + 3x_2 \leq 96 \\ \quad x_1, x_2 \geq 0 \end{array}\right\} \tag{7.48}$$

where $\bar{40} \sim N(40, 15^2)$, $\bar{15} \sim N(15, 8^2)$.

We have $f(\underline{x}) = \bar{40}x_1 + \bar{15}x_2$, the objective function as a normally distributed random variable with *mean* $= 40x_1 + 15x_2$ and standard deviation $\sqrt{225x_1^2 + 64x_2^2}$. Thus, the deterministic equivalent of the given LPP (7.48) is

$$\left.\begin{array}{l} \max \ f(\underline{x}) = k_1(\bar{40}x_1 + \bar{15}x_2) - k_2\sqrt{225x_1^2 + 64x_2^2} \\ \text{subject to } 6x_1 + 5x_2 \leq 420 \\ \quad 4x_1 + 9x_2 \leq 410 \\ \quad 2x_1 + 3x_2 \leq 96 \\ \quad x_1, x_2 \geq 0 \end{array}\right\} \tag{7.49}$$

where k_1 and $k_2 \geq 0$ are selected according to relative importance of mean and standard deviation of the objective function.

Note: If variations in $f(\underline{x})$ are not at all important then $k_1 = 1$ and $k_2 = 0$ in which case the deterministic objective function will reduce to $f(\underline{x}) = 40x_1 + 15x_2$ only and the resulting problem will become a simple LPP.

Let mean and standard deviation have equal importance, then we have $k_1 = 0.5$ and $k_2 = 0.5$. Now the equivalent deterministic problem (7.49) can be solved using optimization software LINGO-18 as:

LINGO-18 codes for the problem (7.49)

MODEL:
MAX= 0.5 * (40 * X1 + 15 * X2) - 0.5 * (225 * X1 ^ 2 + 64 * X2 ^ 2)
^ (1 / 2);
6 * X1 + 5 * X2 <= 420;
4 * X1 + 9 * X2 <= 410;
2 * X1 + 3 * X2 <= 96;
END

The local optimal solution derived through LINGO-18 software is $x_1 = 48$, $x_2 = 0$.

Example 114 *Consider a stochastic LPP:*

$$\left.\begin{array}{r} \max\ Z = 60x_1 + 100x_2 \\ \text{subject to}\ P[x_1 + 2x_2 \leq b_1] \geq 0.95 \\ P[2x_1 + 3x_2 \leq b_2] \geq 0.95 \\ P[x_1 + 4x_2 \leq b_3] \geq 0.95 \\ x_1, x_2 \geq 0 \end{array}\right\} \quad (7.50)$$

where b_i; $i = 1, 2, 3$ are independent normal random variable as $b_1 \sim N(50, 20^2)$, $b_2 \sim N(60, 25^2)$, and $b_3 \sim N(40, 15^2)$. It can be seen from the standard normal table that $P(Z \geq Z_i) = 0.95$, when $Z_i = -1.64$, $i = 1, 2, 3$.

Thus, we have the equivalent deterministic constraints as

$$x_1 + 2x_2 \leq \bar{b}_1 + Z_\alpha \sigma_{b_1}$$

or $x_1 + 2x_2 \leq 50 + (-1.64) \times 20$

or $x_1 + 2x_2 \leq 50 - 32.8$

or $x_1 + 2x_2 \leq 17.2$

Similarly, the second and third constraints become

$$2x_1 + 3x_2 \leq 19$$

$$x_1 + 4x_2 \leq 15.4$$

The final deterministic problem is

$$\begin{aligned} \max \ Z &= 60x_1 + 100x_2 \\ \text{subject to } x_1 + 2x_2 &\leq 17.2 \\ 2x_1 + 3x_2 &\leq 19 \\ x_1 + 4x_2 &\leq 15.4 \\ x_1, x_2 &\geq 0 \end{aligned} \quad (7.51)$$

Now, the equivalent deterministic problem (7.51) can be solved using optimization software LINGO-18 as:

LINGO-18 codes for the problem (7.51)
MODEL:
MAX= 60 * X1 + 100 * X2;
X1 + 2 * X2 <= 17.2;
2 * X1 + 3 * X2 <= 19;
X1 + 4 * X2 <= 15.4;
END

The global optimal solution derived through LINGO-18 software is $x_1 = 5.96$, $x_2 = 2.36$ with max $Z = 593.6$.

Example 115 *Consider the chance constraint stochastic LPP:*

$$\begin{aligned} \max \ Z &= 50x_1 + 120x_2 \\ \text{subject to } P[\overline{10}x_1 + \overline{5}x_2 &\leq 2500] \geq 0.99 \\ P[\overline{4}x_1 + \overline{10}x_2 &\leq 2000] \geq 0.99 \\ P[\overline{1}x_1 + \overline{1.5}x_2 &\leq 450] \geq 0.99 \\ x_1, x_2 &\geq 0 \end{aligned} \quad (7.52)$$

where $\overline{10} \sim N(10, 6^2)$, $\overline{5} \sim N(5, 4^2)$, $\overline{4} \sim N(4, 4^2)$, $\overline{10} \sim N(10, 7^2)$, $\overline{1} \sim N(1, 2^2)$, $\overline{1.5} \sim N(1.5, 3^2)$ and all the random variables are independently normally distributed.

Since $Z_{0.99} = 2.33$, thus the deterministic constraints are as follows:

$$10x_1 + 5x_2 + 2.33\sqrt{36x_1^2 + 16x_2^2} \leq 2500$$

$$4x_1 + 10x_2 + 2.33\sqrt{16x_1^2 + 49x_2^2} \leq 2000$$

and $$x_1 + 1.5x_2 + 2.33\sqrt{4x_1^2 + 9x_2^2} \leq 450$$

Thus, the deterministic equivalent problem is

$$\begin{aligned}
\max \quad & Z = 50x_1 + 120x_2 \\
\text{subject to} \quad & 10x_1 + 5x_2 + 2.33\sqrt{36x_1^2 + 16x_2^2} \leq 2500 \\
& 4x_1 + 10x_2 + 2.33\sqrt{16x_1^2 + 49x_2^2} \leq 2000 \\
& x_1 + 1.5x_2 + 2.33\sqrt{4x_1^2 + 9x_2^2} \leq 450 \\
& x_1, x_2 \geq 0
\end{aligned} \qquad (7.53)$$

Now, the equivalent deterministic problem (7.53) can be solved using optimization software LINGO-18 as:

LINGO-18 codes for the problem (7.53)

MODEL:
MAX= 50 * X1 + 120 * X2;
10 * X1 + 5 * X2 + 2.33 * (36 * X1 ^ 2 + 16 * X2 ^ 2) ^ (1 / 2) <= 2500;
4 * X1 + 10 * X2 + 2.33 * (16 * X1 ^ 2 + 49 * X2 ^ 2) ^ (1 / 2) <= 2000;
X1 + 1.5 * X2 + 2.33 * (4 * X1 ^ 2 + 9 * X2 ^ 2) ^ (1 / 2) <= 450;
END

The local optimal solution derived through LINGO-18 software is $x_1 = 35.24$, $x_2 = 44.01$ with max $Z = 7043.63$.

7.8 Interval Optimization Problem

7.8.1 Introduction

As discussed in Section (7.1), in solving practical optimization problems, it is very difficult for the decision-makers to estimate the precise values of the parameters under study. Such uncertainty in the parameters can be represented as bounded interval numbers and the optimization problems where the parameters either in the objective function or constraints or in both are represented as bounded interval numbers is known as Interval Optimization or an Interval Programming Problem (IPP).

7.8.2 Interval Arithmetic

An interval number consists of both lower and upper bounds, i.e., $\tilde{y} = [y^l, y^u]$, where $y^l \leq y^u$. In this section, some basic arithmetic operations of interval numbers are given:

Let $\tilde{y}_1 = (y_1^l, y_1^u)$ and $\tilde{y}_2 = (y_2^l, y_2^u)$ be two interval numbers, then

(i) Addition: The addition of two interval numbers can be done as:
$$\tilde{y}_1 + \tilde{y}_2 = (y_1^l + y_2^l, y_1^u + y_2^u)$$

(ii) Subtraction: The two interval numbers can be subtracted as:
$$\tilde{y}_1 - \tilde{y}_2 = (y_1^l - y_2^l, y_1^u - y_2^u)$$

(iii) Multiplication: The two interval numbers can be multiplied as:
$$\tilde{y}_1 \times \tilde{y}_2 = [min(y_1^l y_2^l, y_1^l y_2^u, y_1^u y_2^l, y_1^u y_2^u), max(y_1^l y_2^l, y_1^l y_2^u, y_1^u y_2^l, y_1^u y_2^u)]$$

(iv) Division: The two interval numbers can be divided as:
$$\tilde{y}_1 \div \tilde{y}_2 = \{y_1^l \times y_1^u\} \times \left\{\frac{1}{y_2^u}, \frac{1}{y_2^l}\right\}$$

(v) Center and Width: If $\tilde{y} \in [y^l, y^u]$ is a grey or interval number, then the center y_c and width y_w is defined as follows:
$$y_c = \frac{1}{2}(y^l + y^u)$$
$$y_w = \frac{1}{2}(y^u - y^l)$$

(v) Scalar Multiplication: Let k be a scalar, then
$$k\tilde{y} = [ky^l, ky^u], \text{ if } k \geq 0$$
$$k\tilde{y} = [ky^u, ky^l], \text{ if } k < 0$$

7.8.3 Formulation of Interval Optimization Problem

In a linear optimization problem, when the values of parameters in the objective function or constraints or both cannot be precisely estimated, in that case if the upper and lower limits of the parameters or coefficients are known than it can be formulated as an Interval Optimization Problem (IOP).

The mathematical model of an IOP is as follows:

Case I: When all the parameters c_j, y_j, a_{ij}, and b_i are interval numbers.

$$\left.\begin{aligned} \max(\min) \ Z &= \sum_{j=1}^{n} \tilde{c}_j \tilde{y}_j \\ \text{subject to} \ &\sum_{j=1}^{n} \tilde{a}_{ij} \tilde{y}_j \{\leq, =, \geq\} \tilde{b}_i; \ i = 1, 2, \ldots, m \\ \text{and} \ &\tilde{y}_j \geq 0, \ j = 1, 2, \ldots, n \end{aligned}\right\} \quad (7.54)$$

where

$\tilde{y}_j \in (y_j^l, y_j^u) \to$ decision variables in interval form.
$\tilde{c}_j \in (c_j^l, c_j^u) \to$ objective function coefficients in interval form.
$\tilde{a}_{ij} \in (a_{ij}^l, a_{ij}^u) \to$ constraint coefficients in interval form.
$\tilde{b}_i \in (b_i^l, b_i^u) \to$ right hand side parameters in interval form.

Case II: When the parameters c_j, y_j, and a_{ij} are interval numbers.

$$\left. \begin{array}{l} \max(\min) \ Z = \sum_{j=1}^{n} \tilde{c}_j \tilde{y}_j \\ \text{subject to } \sum_{j=1}^{n} \tilde{a}_{ij} \tilde{y}_j \{\leq, =, \geq\} b_i; \ i = 1, 2, \ldots, m \\ \text{and } \tilde{y}_j \geq 0, \ j = 1, 2, \ldots, n \end{array} \right\} \quad (7.55)$$

Case III: When the parameters c_j and a_{ij} are interval numbers.

$$\left. \begin{array}{l} \max(\min) \ Z = \sum_{j=1}^{n} \tilde{c}_j y_j \\ \text{subject to } \sum_{j=1}^{n} \tilde{a}_{ij} y_j \{\leq, =, \geq\} b_i; \ i = 1, 2, \ldots, m \\ \text{and } \tilde{y}_j \geq 0, \ j = 1, 2, \ldots, n \end{array} \right\} \quad (7.56)$$

Case IV: When the parameters c_j are interval numbers.

$$\left. \begin{array}{l} \max(\min) \ Z = \sum_{j=1}^{n} \tilde{c}_j y_j \\ \text{subject to } \sum_{j=1}^{n} a_{ij} y_j \{\leq, =, \geq\} b_i; \ i = 1, 2, \ldots, m \\ \text{and } \tilde{y}_j \geq 0, \ j = 1, 2, \ldots, n \end{array} \right\} \quad (7.57)$$

7.8.4 Algorithm

To solve an IOP follow the following algorithms:

Step 1 First divide the problem into two parts. One with an upper bound and the other with a lower bound.

Step 2 Solve the upper bound problem individually to drive the upper bound (say y_j^*).

Step 3 Now solve the lower bound problem with an additional restriction of the upper bound constraint (say $y'_j \leq y^*_j \ \forall \ j$), $y'^*_j \leq y^*_j \ \forall \ j$.

Step 4 Using the upper and lower bounds, convert the objective function into goals by introducing under and over deviational variables, i.e., $d^-_L, d^+_L, d^-_U, d^+_U$.

Step 5 Finally, construct the model by minimizing the unwanted deviational variables of the goals as:

$$\left.\begin{array}{l} \min \ d^-_L + d^+_U \\ \text{subject to} \ \sum_{j=1}^{n} c^u_j y_j + d^-_L - d^+_L = Z^l \\ \qquad \sum_{j=1}^{n} c^l_j y_j + d^-_U - d^+_U = Z^u \\ \qquad \sum_{j=1}^{n} a_{ij} y_j \{\leq, =, \geq\} b_i; \ i = 1, 2, \ldots, m \\ \text{and} \ \tilde{y}_j \geq 0, \ j = 1, 2, \ldots, n \end{array}\right\} \quad (7.58)$$

7.8.5 Numerical Example using LINGO-18

In this section a numerical example has been presented to demonstrate the algorithm of interval optimization.

Example 116 *Consider a linear optimization problem in which the objective function coefficients are in interval numbers. Find the optimal solution of the given interval optimization problem.*

$$\left.\begin{array}{l} \max \ Z(\tilde{y}) = [15, 17]y_1 + [15, 20]y_2 + [10, 30]y_3 \\ \text{subject to} \ y_1 + y_2 + y_3 \leq 30 \\ \qquad y_1 + 2y_2 + y_3 \leq 40 \\ \qquad y_1 + 4y_3 \leq 60 \\ \text{and} \ \tilde{y}_j \geq 0, \ j = 1, 2, 3 \end{array}\right\} \quad (7.59)$$

Following the step 1 of algorithm the problem (7.59) divided into upper bound and lower bound problem as:

Upper Bound Problem:

$$\left.\begin{array}{rl} \max \ Z(\tilde{y}) = & 17y_1 + 20y_2 + 30y_3 \\ \text{subject to} \ & y_1 + y_2 + y_3 \leq 30 \\ & y_1 + 2y_2 + y_3 \leq 40 \\ & y_1 + 4y_3 \leq 60 \\ \text{and} \ & \tilde{y}_j \geq 0, \ j = 1,2,3 \end{array}\right\} \quad (7.60)$$

Lower Bound Problem:

$$\left.\begin{array}{rl} \max \ Z(\tilde{y}) = & 15y_1 + 15y_2 + 10y_3 \\ \text{subject to} \ & y_1 + y_2 + y_3 \leq 30 \\ & y_1 + 2y_2 + y_3 \leq 40 \\ & y_1 + 4y_3 \leq 60 \\ \text{and} \ & \tilde{y}_j \geq 0, \ j = 1,2,3 \end{array}\right\} \quad (7.61)$$

Firstly, to solve the problem (7.60) for obtaining the worst upper bound, the direct LINGO-18 codes are as:

```
MODEL:
MAX= 17 * Y1 + 20 * Y2 + 30 * Y3;
Y1 + Y2 + Y3 <= 30;
Y1 + 2 * Y2 + Y3 <= 40;
Y1 + 4 * Y3 <= 60;
END
```

The resultant worst upper bound is $y_1 = 6.67, y_2 = 10$ and $y_3 = 13.33$ with $Z^u = 713.33$.

Solving the problem (7.61) to obtain the best lower bound, the direct LINGO-18 codes are as:

```
MODEL:
MAX= 15 * Y1 + 15 * Y2 + 10 * Y3;
Y1 + Y2 + Y3 <= 30;
Y1 + 2 * Y2 + Y3 <= 40;
Y1 + 4 * Y3 <= 60;
Y1 <= 6.67;
Y2 <= 10;
Y3 <= 13.33;
END
```

The resultant best lower bound is $y_1 = 6.67, y_2 = 10$ and $y_3 = 13.33$ with $Z^l = 383.35$.

Now, using the obtained upper and lower bounds the objective function can be converted into goals by introducing under and over deviational variables and the goal model can be written as:

Direct LINGO-18 codes for Goal Model

MODEL:
MIN= DLNG + DUPO;
17 * Y1 + 20 * Y2 + 30 * Y3 + DLNG - DLPO = 383.35;
15 * Y1 + 15 * Y2 + 10 * Y3 + DUNG - DUPO = 713.33;
Y1 + Y2 + Y3 <= 30;
Y1 + 2 * Y2 + Y3 <= 40;
Y1 + 4 * Y3 <= 60;
END

The resultant optimal solution of the given interval optimization problem (7.59) is $y_1 = 30, y_2 = 0,$ & $y_3 = 0$. Now using the interval arithmetic operations the value of the objective function is:

$$Z(\tilde{y}) = [15, 17] \times 30 + [15, 20] \times 0 + [10, 30] \times 0$$

$$Z(\tilde{y}) = [450, 510] + [0, 0] + [0, 0]$$

$$Z(\tilde{y}) = [450, 510]$$

Chapter 8
Multi-Objective Optimization

8.1 Introduction

Decision-making is a complicated task that policy-makers need to complete with the desired outcome. *Optimization* is merely obtaining the best result out of many possible alternatives. In reality, it is very seldom that we have a problem with only one solution. However, the various solution alternatives must be carefully investigated before arriving at a final solution scientifically. Most of the managerial decision-making problems involve more than one goal with multiple alternatives, which must be achieved under limited resources such as time, budget, space, and others. In practice, these objectives will be conflicting; they cannot reach their optimal values simultaneously. If they could, then the model can be solved as a single-objective problem for any of the objectives. The space formed by the values of the set of objectives is known as objective space. Fig. (8.3) shows an example of an objective space for a bi-objective decision problem. A decision problem with more than one objective (maximized or minimized) is called a *multi-objective optimization problem* (MOOP).

A MOOP with inequality can be written mathematically as:

$$\left. \begin{array}{l} \text{Find } \mathbf{X} = \begin{pmatrix} x_1 \\ \vdots \\ x_n \end{pmatrix} \\ \min \ f(x) = [f_1(\mathbf{X}), f_2(\mathbf{X}), \cdots, f_k(\mathbf{X})] \\ \text{subject to } g_j(\mathbf{X}) \leq b, \ j = 1, 2, \cdots, m \end{array} \right\} \quad (8.1)$$

where k denotes the number of conflicting objective functions to be minimized, $g_j(\mathbf{X}) \leq b$ representing m constraint equations, and x is an $n-dimentional$ vector of decision variables in $X = \{x \in \mathbb{R}^n\}$.

Notes:

- Equality constraints if exist, can also be included in the formulation of Eqn. (8.1).
- Any or all of the functions $f(\mathbf{X})$ and $g_i(\mathbf{X})$ may be non-linear.
- The MOOP is also known as the vector optimization problem.

Some of the basic definitions related to MOOP are as follows:

Definition 8.1 A *criterion* is a single measure by which the goodness of any solution to a decision problem can be measured. Many possible criteria are arising from different application fields. Still, some of the most commonly arising ones relate to the highest level of cost, profit, time, distance, performance of a system, company or organizational strategy, and personal preferences of the decision-maker(s), safety considerations.

Definition 8.2 A decision problem with more than one criterion is known as multi-criteria decision making (MCDM) problem; sometimes, it is called multi-criteria decision analysis (MCDA).

Definition 8.3 An *objective* is a criterion with the additional directional information (maximize or minimize) in which the decision-maker(s) prefers on the criterion scale. For example, minimize cost or maximize the performance of a system.

Definition 8.4 A *goal* is referred to as a criterion and a numerical level, known as a target level, which the decision-maker(s) desires to achieve on that criterion. There are three principal types of goals that can occur in a goal programming model.

Goal 1 Achieving at most the targeted value, for example, keeping the cost within a budgeted limit of $100.

Goal 2 Achieving at least the targeted value, for example, aims to produce at least 200 products.

Goal 3 Achieving the targeted value exactly, for example, aims to employ exactly 200 specialists.

Definition 8.5 A *deviational variable* measures the difference between the target level on a criterion and the value achieved in a given solution. If the achieved value is above the target level, then the difference is given by the value of the *positive deviational variable*. If the achieved value is below the target level, then the difference is given by the value of the *negative deviational variable*.

Definition 8.6 *Pareto-Efficient Solution*—A solution to a multi-objective problem is Pareto efficient if no other feasible solution exists that is at least as good for all objectives and strictly better for at least one objective. It is also known as *Pareto optimal* or in objective space as *non-dominated*.

Definition 8.7 *Pareto-Inefficient Solution*—A solution to a multi-objective problem is *Pareto inefficient* if another feasible solution exists that is at least as good for all objectives and strictly better for at least one objective. It is also known as *Pareto suboptimal* or in objective space as *dominated*.

Definition 8.8 *Goal Balancing*—This is a situation of balancing between under and over achievement of two goals. Goals should be balanced rather than mere achievement.

8.1.1 Pareto Optimal Solution

In general, no solution vector **X** exists that optimizes all the k objective functions simultaneously. Hence, a new concept known as *pareto optimum solution* is used in MOOP.

Definition 8.9 A feasible vector **X** is called a *strict efficient solution* or a *strict Pareto optimum* for MOOP if there exists no other feasible solution **Y** such that $f_i(\mathbf{Y}) \leq f_i(\mathbf{X})$, $for\ i = 1, 2, \cdots, k$ with at least one strict inequality, i.e., $f_j(\mathbf{Y}) < f_j(\mathbf{X})$ for at least one j. In other words, a feasible vector **X** is called a *strict Pareto optimum* if there is no other feasible solution **Y** that would reduce some objective function without causing a simultaneous increase in at least one other objective function. They are also called *ideal solution, superior solution, or utopia point*.

Definition 8.10 A feasible vector **X** is said to be a *weak efficient solution* or a *weak Pareto optimum* for MOOP if and only if there is no other feasible solution **Y** such that $f_i(\mathbf{Y}) < f_i(\mathbf{X})$ for all $i \in \{1, 2, \cdots, n\}$.

For example, points p_1 and p_5 are weak Pareto optima; points p_2, p_3 and p_4 are strict Pareto optima (see Fig. (8.1)).

Another example is if two objective functions are given by $f_1(x-3)^4$ and $f_2(x-2)^6$, all the values of x between 3 & 6 (points on the line segment PQ) on their graphs are *pareto optimum solutions* (see Fig. (8.2)).

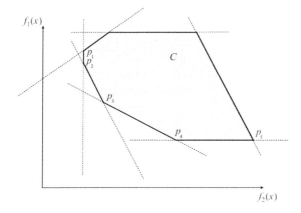

Figure 8.1: Weak and Strict Pareto Optima.

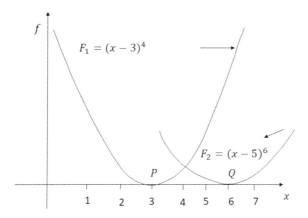

Figure 8.2: Pareto Optimal Solutions.

8.1.2 Ideal Point

Definition 8.11 The point in objective space at which each objective in a multi-objective optimization problem takes its optimal value when optimized individually, within the feasible region, is known as the *ideal point*. If the objectives are conflicting, then this point will be outside the feasible region in objective space and hence an *infeasible point*. Nevertheless, it provides a useful point of reference to measure the goodness of any solution.

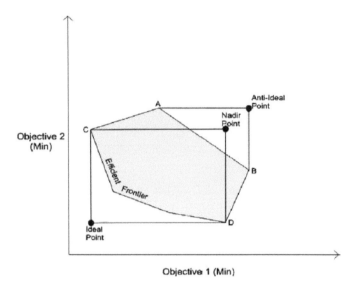

Figure 8.3: Pareto Optimal Solution.

8.1.3 Anti-ideal Point

Definition 8.12   Anti-ideal is the point in objective space formed from the anti-optimization (i.e., maximize an objective the decision-maker(s) wish to minimize or minimize an objective the decision-maker(s) wish to maximize) of each objective within the feasible region. To provide correct scaling, a measure of the worst solution and the best for each objective is needed. The anti-ideal point is one possibility for this measure.

8.1.4 Nadir Point

Definition 8.13   This is defined as the worst value from amongst the Pareto frontier for each objective. The major drawback of the anti-ideal point is that it is calculated from solutions that are not Pareto efficient (e.g., from points A and B in Fig. (8.3)). This means that the decision-making axiom known as 'independence from irrelevant alternatives' is broken. In theory, a better solution would be to use the nadir point, which is defined as the worst value from amongst the Pareto frontier for each objective. The nadir point can be seen in Fig. (8.3) as being calculated from solutions C and D, which are both Pareto efficient.

8.1.5 Pareto Frontier

Definition 8.14 The set of all Pareto-efficient solutions to a decision problem is known as the Pareto frontier. It is also known as *Pareto front, Pareto set,* and in objective space as the *non-dominated set*. An example of *Pareto front* is shown in Fig. (8.4) defined by the points between $(f_2(\hat{x}), f_1(\hat{x}))$ and $(f_1(\tilde{x}), f_2(\tilde{x}))$. These points are known as non-dominated or non-inferior points in MOOP.

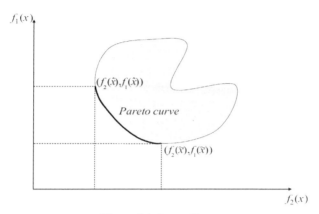

Figure 8.4: Pareto Front.

8.2 Multi-Objective Optimization Techniques

Several techniques have been developed over the years to solve the MOOP, most of which generate a set of *pareto optimum solutions* and use some additional rules or criterion to select one particular *pareto optimum solution* as the best solution for MOOP. Some of the best techniques are discussed in the following sections.

8.2.1 Weighted Technique

This method combines the MOOP into a single objective using a scalar function. It is also known as *weighted-sum technique*. It minimizes a positively weighted convex sum of the objectives, representing a new optimization problem with a unique objective function. In the weighting method, MOOP is formulated by taking the weighted sum of all the objective functions, i.e.,

$$\min \ Z(x) = \sum_{i=1}^{k} w_i f_i(x)$$
$$\text{subject to} \ x \in X$$
$$\sum w_i = 1, \ w \geq 0 \quad (8.2)$$

where $w_i = [w_1, w_2, \ldots, w_k]$ is the weighted coefficient vector assigned to the k objective functions.

8.2.2 Goal Programming Technique

Goal Programming (GP) is a multi-objective optimization technique used to solve a decision-making problem with nested conflicting goals, with target value(s) to attain a certain aspirational level. The minimization of unwanted deviations involve, using an appropriate model function. In GP approach, the Decision Maker is normally confronted with different goals that are conflicting in nature, say f_i ($i = 1, 2, 3, \ldots, p$) with different conflicting criteria and some level of acceptable aspiration for their goals g_i, ($i = 1, 2, 3, \ldots, p$) to be achieved within a feasible set, say $D = h_s(x) \leq 0, s = 1, 2, 3, \ldots m$. Generally, GP formulation is used to minimize deviations arising from over-achievement and under-achievement of the decision-makers particular goal. There is a different model of GP depending on the nature of the DM's problem, goals, and aspirations.

A typical GP model that minimizes the deviational variables is given as:

$$\min \sum_{k=1}^{K} |F_k(X) - g_k|$$
$$\text{subject to } \sum_{j=1}^{n} a_{ij} x_j (\leq, =, \geq) b_i, \ i = 1, 2, \ldots, m, \quad (8.3)$$
$$x_j \geq 0, \ j = 1, 2, \ldots, n$$

In Eqn. (8.3), K is the total goals, $F_k(X)$ is the kth objective, g_k is the kth aspirational goal, a_{ij} is the technological coefficient, and b_i the available resources or limitation for the ith constraint. In the simplest standard form Eqn. (8.3) can be written as:

$$\min \sum_{k=1}^{K} (\eta_k + \rho_k)$$
$$\text{subject to } F_k(X) + \eta_k - \rho_k = g_k, \ k = 1, 2, \ldots, K,$$
$$\sum_{j=1}^{n} a_{ij} x_j (\leq, =, \geq) b_i, \ i = 1, 2, \ldots, m, \quad (8.4)$$
$$\eta_k, \rho_k, x_j \geq 0, \ k = 1, 2, \ldots, K, \ j = 1, 2, \ldots, n$$

where $\eta_k = \max\{0, F_k(X) - g_k\}$ and $\rho_k = \max\{0, g_k - F_k(X)\}$ are the over and under achievement values for the kth aspirational goal respectively.

The essence of goal programming is the minimization of unwanted deviational variables. For goal type 1 or 'less is better', the positive deviational variable is said to be the unwanted deviational variable. For goal type 2 or 'more is better', the negative deviational variable is said to be the unwanted deviational variable. For goal type 3, both positive and negative deviational variables are said to be unwanted deviational variables.

In GP, the ideal and anti-ideal point is often used to calculate the normalization constants when using zero–one normalization, an example of both ideal and anti-ideal point is given in Fig. (8.3).

In many GP problems, it is insufficient to consider the average level of achievement of the goals solely without looking at the balance between achieving the goals. To illustrate this fact, Table 8.1 shows two solutions to a simple two-goal programme where negative deviations from the target level are unwanted. The two solutions have the same average level of achievement but different levels of balance between goals. Clearly, if balancing the achievement of the goals is of any importance to the decision maker then they will prefer solution 2 over solution 1. However, if only the average level of achievement is considered when solving the goal programme then solutions 1 and 2 are shown to have the same value to the decision maker. Goal programming contains elements of optimization. There are three situations to note where the optimizing philosophy has particular importance:

Table 8.1: Example of Two-Goal Model.

Solutions	Goal 1		Goal 2	
	Target	Achieved	Target	Achieved
Solution 1	250	230	100	90
Solution 2	250	250	100	80

1. If the goal target levels are set very optimistically, such as at their ideal values, in the goal programme, then the dominant underlying philosophy has changed from satisfying to optimizing.

2. If Pareto optimality detection and restoration occur, then the goal programme has a mix of satisfying and optimizing philosophies.

3. If the goals are two-sided (i.e., a particular value is optimal rather than a 'more is better' or 'less is better' situation) then the satisfying and optimizing philosophies can be thought of as coinciding for those goals.

8.2.2.1 Algorithm of Goal Programming Technique

The following are the step-wise algorithm for solving the GP model.

Step 1 Formulate the real-life situation as a multiple objective linear/non-linear programming problem.

Step 2 Make the individual objective functions with their goal values as additional constraints of the problem.

Step 3 Standardize the constraints in Step 2 with under and over achievement deviational variables as the case might be.

Step 4 Convert the multi-objective problem into a single problem by minimizing the sum of deviational variables subject to the goal equations (Step 2), and the functional constraints.

Step 5 Finally, solve the model developed in Step 4 using any available computer software such as LINDO/LINGO, GAMS, and more.

8.2.2.2 Another Standard Form of GP

In this method decision-makers set some target values to achieve, and the technique tries to minimize the under and over achievements of these targets. The mathematical formulation of GP for MOOP can also be written as:

$$\min \ Z = \sum_{j=1}^{k} \left[(d_j^+ + d_j^-)^p\right]^{1/p}, \ p \geq 1,$$
$$\text{subject to} \ g_j(\underline{x}) \leq 0, \ \forall j \in \{1, 2, \cdots, m\},$$
$$f_j(\underline{x}) + d_j^+ - d_j^- = b_j, \forall j \in \{1, 2, \cdots, k\},$$
$$d_j^+, d_j^- \geq 0, \forall j$$
$$d_j^+ \cdot d_j^- = 0, \forall j$$
(8.5)

where b_j is the goal set by DM for the jth objective, and d_j^+, d_j^- are the under and over achievements of the jth goal. The value of p is chosen by the DM based on the utility function.

Note that for $p = 1$ in Eqn. (8.5), all deviations from the goal values are taken into account in direct proportion to their magnitudes. For $p = \infty$, only the largest deviation has the greatest influence and hence taken into account. Mathematically,

$$\min \ Z = \sum_{j=1}^{k} (d_j^+ + d_j^-),$$
$$\text{subject to} \ g_j(\underline{x}) \leq 0, \ \forall j \in \{1, 2, \cdots, m\},$$
$$f_j(\underline{x}) + d_j^+ - d_j^- = b_j, \forall j \in \{1, 2, \cdots, k\},$$
$$d_j^+, d_j^- \geq 0, \forall j$$
$$d_j^+ \cdot d_j^- = 0, \forall j$$
(8.6)

A typical goal for the jth objective b_j, is found by first solving Eqn. (8.7) given as:

$$\min \ f_j(\underline{x})$$
$$\text{subject to} \ g_j(\underline{x}) \leq 0, \ j = 1, 2, \cdots, m$$
(8.7)

If the solution of Eqn. (8.7) is $\underline{x}_j^*$, then b_j is taken to be $b_j = f_j(\underline{x}^*)$.

Many approaches for GP techniques exist; these include the lexicographical, the interactive GP, the reference GP and the multi-criteria GP amongst others.

8.2.3 Lexicographic Goal Programming Technique

In lexicographic GP, the objectives or goals are ranked in order of importance by the DM, this can be done by ordering the unwanted deviations into different levels of priority, with the minimization of a deviation in a higher priority level being infinitely more important than any deviation in lower priority level. The objectives can be divided into different priority classes and no two goals will have equal priority. The goals are given ordinal ranking and are called preemptive factors. The priority relationships imply that multiplication by n, however large it may be, cannot make the lower level goal as the higher goal, i.e., $P_i > P_{i+1}$. Mathematically, the model can be stated as:

$$\max \ Z = \sum_{i=1}^{m} P_i(\delta_i^+ + \delta_i^-)$$

$$\text{subject to} \ \sum_{j=1}^{n} a_{ij}x_j + \delta_i^- - \delta_i^+ = b_i, \ i = 1, 2, 3, \ldots, m \quad (8.8)$$

$$\sum_{j=1}^{n} a_{ij}x_j (\leq, =, \geq) \ b_i \ i = m+1, m+2, \ldots, m+p$$

$$\delta_i^+, \delta_i^-, x_j \geq 0, \ i = 1, 2, \ldots, m, j = 1, 2, \ldots, n$$

where there are m goals, p system constraints, k priority levels and n decision variables. P_i is the preemptive priority factors of the ith goal.

In other words, after ranking the objectives according to DM's priority, the optimal solution $(\underline{x}^*)$ is then found by minimizing the objective functions starting with the most important and proceeding according to the ranking order of objectives.

Let $f_1(\underline{x})$ and $f_k(\underline{x})$ denote the most and least important objective functions respectively. The first problem is formulated as:

$$\min \ f_1(\underline{x})$$
$$\text{subject to} \ g_j(\underline{x}) \leq 0, \ j = 1, 2, \ldots, m \quad (8.9)$$

Then, the solution vector of Eqn. (8.9) is $\underline{x}_1^*$ and $f_1^* = f_1(\underline{x}_1^*)$ is obtained. Similarly, the second problem can be formulated as:

$$\min \ f_2(\underline{x}) + \delta_1$$
$$\text{subject to} \ g_j(\underline{x}) \leq 0, \ j = 1, 2, \ldots, m \quad (8.10)$$
$$f_1(\underline{x}) - \delta_1 = f_1^*$$

The solution vector of Eqn. (8.10) is obtained as $f_2^* = f_2(\underline{x}_2^*)$. This procedure is repeated until all the k objectives have been exhausted. The ith problem is given by

$$\min \ f_i(\underline{x}) + \sum_{i=1}^{i-l} \delta_i$$

$$\text{subject to } g_j(\underline{x}) \leq 0, \ j = 1, 2, \ldots, m$$

$$f_l(\underline{x}) - \delta_l = f_l^*, \ l = 1, 2, \ldots, i-1 \quad (8.11)$$

and the solution of Eqn. (8.11) is obtained as $\underline{x}_i^*$ and $f_i(\underline{x}_i^*)$. Finally, the solution obtained at the end (i.e., x_k^*) is taken as the desired solution $\underline{x}^*$ for the original MOOP.

8.2.4 ε-Constraints Technique

In this technique, the DM chooses only one objective out of n objective functions to minimize, considering the remaining as constraints less than or equal to the target values. It was proposed by [34]. Mathematically, it can be represented as:

$$\min \ f_j(x)$$
$$f_i(x) \leq \varepsilon_i, \ \forall \ i \in \{1, 2, \cdots, n\} \setminus \{j\} \quad (8.12)$$
$$x \in S$$

where f_j is the chosen objective, S the feasible solution space and ε_i the target values for various objectives. The solution x^* to Eqn. (8.12) is a *weak Pareto optimum*.

8.2.5 Fuzzy Goal Programming Technique

Combination of fuzzy set theory concept and GP results in fuzzy goal programming (FGP) technique. FGP deals with imprecise goals of a decision-maker. These goals can be associated with an objective function or the constraints, and there value or degree of association ranges from zero to one. FGP allows the decision-maker who cannot define goals in a precise manner to at least express them in weighting structure. A generalized model for such type of problems (FGP) is:

$$\text{Find } X = (x_1, x_2, \ldots, x_n),$$
$$\text{such that } Z_k(X)(\succeq, \simeq, \preceq)\tilde{g}_k, k = 1, 2, 3, \ldots, K$$
$$AX \leq b_i, i = 1, 2, \ldots, m \quad (8.13)$$
$$X \geq 0$$

where g_k is the vector of individual goals, b_i is the vector of individual resources available, and A is the technological coefficient. The symbol $\succeq$ is for maximization goal type, means that the objective should be more than or exactly equal to the aspiration goal g_k. The solution is acceptable by the DM even if it is less than g_k by a certain level. The symbol $\preceq$ stands for minimization goal type, meaning that the objective solution should be less than or exactly equal to the aspiration g_k, it could be accepted up to the allowable limit (tolerance). In contrast, the symbol $\simeq$ stands for equality type goals, implying that the solution should be within the level of aspiration g_k flexibly. The solution can be satisfied by the DM even if it is less than or greater than g_k to a certain level of tolerance. Z_k is the kth fuzzy objective while X is the n-dimensional vector for decision variables.

The membership function for a maximization fuzzy-goal type $Z_k(X) \succeq g_k$ problem is given as:

$$\mu_k(Z_k(X)) = \begin{cases} 1, & if\ Z_k(X) \geq g_k, \\ \dfrac{Z_k(X) - L_k}{g_k - L_k}, & if\ L_k \leq Z_k(X) \leq g_k, \\ 0, & if\ Z_k(X) \leq L_k. \end{cases} \quad (8.14)$$

And, the membership function for a minimization fuzzy-goal type $Z_k(X) \preceq b_k$ problem is given as:

$$\mu_k(Z_k(X)) = \begin{cases} 1, & if\ Z_k(X) \leq g_k, \\ \dfrac{U_k - Z_k(X)}{U_k - g_k}, & if\ g_k \leq Z_k(X) \leq U_k, \\ 0, & if\ Z_k(X) \geq U_k. \end{cases} \quad (8.15)$$

The membership function for the fuzzy-equal goal of type $(Z_k(X) \simeq g_k)$ is given as:

$$\mu_k(Z_k(X)) = \begin{cases} 0, & if\ Z_k(X) \leq L_k \\ \dfrac{Z_k(X) - L_k}{g_k - L_k}, & if\ L_k \leq Z_k(X) \leq g_k \\ \dfrac{U_k - Z_k(X)}{U_k - g_k}, & if\ g_k \leq Z_k(X) \leq U_k \\ 0, & if\ Z_k(X) \geq U_k \end{cases} \quad (8.16)$$

where U_k and L_k are the upper and lower bounds of the g_k for k^{th} goal set by the DM. Now, the final FGP model will be:

$$\left. \begin{array}{l} \max\ \lambda \\ \text{subject to}\ \lambda \leq \mu_k(Z_k(X)),\ k = 1, 2, \ldots, K \\ AX \leq \tilde{b}_i,\ i = 1, 2, \ldots, m \\ \text{and}\ X \geq 0 \end{array} \right\} \quad (8.17)$$

Now, solve it using optimization software LINGO-18. After solving the model, we get the most preferred fuzzy efficient, and compromised solution for MOOP.

8.2.6 Fuzzy Goal Programming with Tolerance

The constraint of a fuzzy model is a subset of vector X with a membership characteristic function $\mu_{a_{ij}}(x_j): x \longrightarrow [0,1]$, given by

$$\mu_{\Sigma a_{ij}x_j \gtrsim b_i} = \begin{cases} 1, & if \ \sum_{j=1}^{n} a_{ij}(x_j) = b_i, \ i=1,2,\ldots,m \\ \dfrac{b_i + T*b_i - \sum_{j=1}^{n} a_{ij}(x_j)}{T*b_i}, & if \ b_i \leq \sum_{j=1}^{n} a_{ij}(x_j) \leq b_i + T*b_i, \ i=1,2,\ldots,m \\ 0, & if \ \sum_{j=1}^{n} a_{ij}(x_j) \geq b_i + T*b_i \end{cases}$$

(8.18)

$$\mu_{\Sigma a_{ij}x_j \lesssim b_i} = \begin{cases} 1, & if \ \sum_{j=1}^{n} a_{ij}(x_j) = b_i, \ i=1,2,\ldots,m \\ \dfrac{\sum_{j=1}^{n} a_{ij}(x_j) - b_i + T*b_i}{T*b_i}, & if \ b_i - T*b_i \leq \sum_{j=1}^{n} a_{ij}(x_j) \leq b_i, \ i=1,2,\ldots,m \\ 0, & if \ \sum_{j=1}^{n} a_{ij}(x_j) \geq b_i \end{cases}$$

(8.19)

The general FGP model with tolerance value is given as:

Find a vector $X = [x_1, x_2, \ldots, x_n]^T$ such that it will

$$\begin{cases} \max \ \lambda \\ \text{subject to} \ \lambda \leq \dfrac{Z_k(x) - l_k}{g_k - l_k}, \ if \ Z_k(x) \gtrsim g_k \\ \lambda \leq \dfrac{U_k - Z_k(x)}{U_k - g_k}, \ if \ Z_k(x) \lesssim g_k \\ \lambda \leq \dfrac{(b_i + T*b_i) - \sum_{i=1}^{m} a_{ij}(x_j)}{T*b_i}, \ if \ \sum_{i=1}^{m} a_{ij}(x_j) \succeq b_i \\ \lambda \leq \dfrac{(\sum_{i=1}^{m} a_{ij}(x_j)) - b_i + T*b_i}{T*b_i}, \ if \ \sum_{i=1}^{m} a_{ij}(x_j) \preceq b_i \\ x_j \geq 0, \ j = 1,2,\ldots,n \\ \lambda \geq 0 \end{cases}$$

(8.20)

8.3 Numerical Example using LINGO-18

Example 117 *For the demonstration of various multi-objective optimization techniques, an example has been taken from [17].*

Manager of outlet sells two types of muffins: Apple and Pineapple. We assume that muffins are so good that there is unlimited demand for them. Everyday, he wants to produce at least 2 batches of apple muffins and always has enough to make 8 batches of pineapple muffins every day. Rest of the data about the parameters are given below in Table (8.2).

Table 8.2: Data for Different Parameters.

Muffins	Employee's Time		Cost	Selling Price	Profit
	X	Y			
Apple	1	2	30	40	10
Pineapple	2	1	45	55	15
Availability	8	7			

Now, the question arises that how many batches of apple and pineapple muffins should be made so that his revenue maximizes and cost minimizes. For that, we have formulated the problem into MOOP. Using the data given in Table (8.2), the MOOP will be:

$$\left. \begin{array}{l} \max \ R = 10x_1 + 15x_2 \\ \min \ C = 30x_1 + 45x_2 \\ \text{subject to} \ x_1 + 2x_2 \leq 8 \\ \qquad 2x_1 + x_2 \leq 7 \\ \qquad x_1 \geq 2 \\ \qquad x_2 \leq 8 \\ \qquad x_i \geq 0; \ i = 1,2 \end{array} \right\} \qquad (8.21)$$

8.3.1 Solution through Weighted Technique

Suppose both the objectives are equally important for the manufacturer, than the weighted sum model will be:

$$\left. \begin{array}{l} \max \ U = 0.5R + 0.5C \\ \text{subject to} \ x_1 + 2x_2 \leq 8 \\ \qquad 2x_1 + x_2 \leq 7 \\ \qquad x_1 \geq 2 \\ \qquad x_2 \leq 8 \\ \qquad x_i \geq 0; \ i = 1,2 \end{array} \right\} \qquad (8.22)$$

Now, the optimum allocation can be derived by solving the problem (8.22) using LINGO-18 software.

> **LINGO-18 codes for the problem** (8.22)
> MODEL:
> MAX= - 20 * X1 - 30 * X2;
> X1 + 2 * X2 <= 8;
> 2 * X1 + X2 <= 7;
> X1 >= 2;
> X2 <= 8;
> END

The global optimal solution is $x_1 = 2$, $x_2 = 0$ with maximum revenue $20, and minimum cost $60.

8.3.2 Solution through Goal Programming Technique

For solving the problem (8.21), the GP model discussed in Section (8.2.2.2) is used. Firstly goals are obtained by solving the problem separately for each objective function as:

> **LINGO-18 codes for the problem with first objective:**
> MODEL:
> MAX= 10 * X1 + 15 * X2;
> X1 + 2 * X2 <= 8;
> 2 * X1 + X2 <= 7;
> X1 >= 2;
> X2 <= 8;
> END

The desired goal for revenue maximization is R=$65.

> **LINGO-18 codes for the problem with second objective:**
> MODEL:
> MIN= 30 * X1 + 45 * X2;
> X1 + 2 * X2 <= 8;
> 2 * X1 + X2 <= 7;
> X1 >= 2;
> X2 <= 8;
> END

The desired goal for cost minimization is C=$60.

Using the desired goals, the goal programming model for the problem (8.21) is:

LINGO-18 codes for the problem (8.21)

MODEL:
MIN= ((D1POS + D1NEQ) ^ 0.5) ^ (1 / 0.5) + ((D2POS + D2NEQ) ^ 0.5) ^ (1 / 0.5);
X1 + 2 * X2 <= 8;
2 * X1 + X2 <= 7;
X1 >= 2;
X2 <= 8;
D1POS - D1NEQ + 10 * X1 + 15 * X2 = 65;
D2POS - D2NEQ + 30 * X1 + 45 * X2 = 60;
END

After solving the problem using LINGO-18, the global optimal solution is $x_1 = 2$, $x_2 = 0$ with maximum revenue \$20, and minimum cost \$60. The results show that the goal for minimizing the cost is attained but the optimal solution for revenue maximization is not close to the goal set by the manufacturer.

8.3.3 Solution through Lexicographic Goal Programming Technique

In this method as discussed in Section (8.2.3) the problem is solved according to the preemptive ranking of the objective functions. First the single objective problem of maximizing revenue is solved as a minimization problem (Max Z=Min (-Z)).

LINGO-18 codes:

MODEL:
MIN= - 10 * X1 - 15 * X2;
X1 + 2 * X2 <= 8;
2 * X1 + X2 <= 7;
X1 >= 2;
X2 <= 8;
END

After solving the problem using LINGO-18, the solution is $R^* = -65$. Now, the next problem will be:

> **LINGO-18 codes:**
> MODEL:
> MIN= 30 * X1 + 45 * X2 + d1;
> X1 + 2 * X2 <= 8;
> 2 * X1 + X2 <= 7;
> X1 >= 2;
> X2 <= 8;
> 10 * X1 + 15 * X2 - d1 = -65;
> END

The solution is $C^* = 60$. Now, the final model to derive the optimal allocation of the problem (8.21) is:

> **LINGO-18 codes:**
> MODEL:
> MIN= D1 + D2;
> X1 + 2 * X2 <= 8;
> 2 * X1 + X2 <= 7;
> X1 >= 2;
> X2 <= 8;
> 10 * X1 + 15 * X2 - D1 = -65;
> 30 * X1 + 45 * X2 - D2 = 60;
> END

The desired optimal allocation is derived using LINGO-18 software as $x_1 = 2$, $x_2 = 0$ with maximum revenue \$20, minimum cost \$60, and deviations $d_1 = 85$, $d_2 = 0$.

8.3.4 Solution through ε-Constraints Technique

Suppose the DM choose to solve the first objective keeping the second in the constraint, then the model to derive the optimal solution of the problem (8.21) is as follows:

> **LINGO-18 codes:**
> MODEL:
> MIN= - 10 * X1 - 15 * X2;
> X1 + 2 * X2 <= 8;
> 2 * X1 + X2 <= 7;
> X1 >= 2;
> X2 <= 8;
> 30 * X1 + 45 * X2 <= 60;
> END

The desired optimal allocation is derived using LINGO-18 software as $x_1 = 2$, $x_2 = 0$. This method provides the weak pareto optimal solution.

8.3.5 Solution through Goal Programming Technique

The GP algorithm discussed in Section (8.2.2.1) has been illustrated through the following example.

Example 118 *An XYZ company wants to produce x chairs and y tables with a unit contribution of $5 and $7 respectively. Each chair and table unit requires 2 & 5 quantities of raw-materials, 3 & 2 man-hours respectively. There is a total of 51 raw-materials and 42 man-hours available for production. The total number of chairs and tables should not exceed 14. Additionally, the manager wants to produce seven tables as a goal. He further states that the total profit would be $100, and there is a penalty of $10 for producing below seven tables and $1 penalty for profit below $100.*

Solution: Following step 1 of the GP algorithm discussed in Section (8.2.2.1), the problem can be formulated as MOOP:

$$\max 5x + 7y = 100, \quad \text{(Goal 1)}$$
$$\max y = 7, \quad \text{(Goal 2)}$$
$$\text{subject to } 2x + 5y \leq 51, \quad \text{(materials constraints)} \quad (8.23)$$
$$3x + 2y \leq 42, \quad \text{(man - hour constraints)}$$
$$x + y \geq 14, \quad \text{(total number of pieces)}$$

Now, after applying step 2 & 3, the final GP model discussed in step 4 of the algorithm is as follows:

$$\min Z = \eta_1 + 10\eta_2$$
$$\text{subject to } 5x + 7y + \eta_1 - \rho_1 = 100, (Goal\,1)$$
$$y + \eta_2 - \rho_2 = 7, (Goal\,2)$$
$$2x + 5y \leq 51, (materials\ constraints) \qquad (8.24)$$
$$3x + 2y \leq 42, (man - hour\ constraints)$$
$$x + y \geq 14, (total\ number\ of\ pieces)$$
$$x, y, \eta_1, \eta_2, \rho_1, \rho_2 \geq 0.$$

The problem (8.24) is solved by an optimization software LINGO-18 and deriving the optimum solution as follows:

LINGO-18 codes for the problem (8.24)

MODEL:
MIN= E1 + 10 * E2;
E1 + 5 * X + 7 * Y - R1 = 100;
E2 + Y - R2 = 7;
2 * X + 5 * Y <= 51;
3 * X + 2 * Y <= 42;
X + Y >= 14;
X >= 0;
Y >= 0;
E1 >= 0;
E2 >= 0;
R1 >= 0;
R2 >= 0;
END

The solution of the model is $x = 8, y = 7\ and\ \eta_1 = 11$. This indicates that DMs goal 2 is fully attained will goal 1 will be left under attained.

8.3.6 Solution through Fuzzy Goal Programming Technique

Example 119 *Let us consider a multi-objective optimization problem as:*

$$\max\ F_1 = 5x_1 - 4x_2$$
$$\max\ F_2 = -2x_1 + 8x_2$$
$$\text{subject to}\ -x_1 + x_2 \leq 16$$
$$x_1 \leq 12 \qquad (8.25)$$
$$x_1 + x_2 \leq 16$$
$$x_2 \leq 8$$
$$x_1, x_2 \geq 0$$

By solving the problem (8.25) for each objective function individually through LINGO-18, the following payoff matrix will be derived.

$$\begin{array}{c} \\ F_1 \\ F_2 \end{array} \begin{pmatrix} F_1 & F_2 \\ 60 & -22 \\ -24 & 60 \end{pmatrix}$$

The lower and upper solutions of F_1 and F_2 can be seen in the payoff matrix above. Since, both the objective functions are of maximization type, the membership function is constructed as follows:

$$\mu_1(F_1(X)) \leq \frac{(5x_1 - 4x_2) - (-24)}{60 - (-24)}, \implies 84\mu_1 \leq (5x_1 - 4x_2) + 24$$

$$\mu_k(F_2(X)) \leq \frac{(-2x_1 + 8x_2) - (-22)}{60 - (-22)}, \implies 82\mu_2 \leq (-2x_1 + 8x_2) + 22$$

Finally, the FGP model is:

$$\max\ \mu_1 + \mu_2$$
$$\text{subject to}\ 84\mu_1 \leq (5x_1 - 4x_2) + 24$$
$$82\mu_2 \leq (-2x_1 + 8x_2) + 22$$
$$-x_1 + x_2 \leq 16$$
$$x_1 \leq 12 \qquad (8.26)$$
$$x_1 + x_2 \leq 16$$
$$x_2 \leq 8$$
$$x_1, x_2 \geq 0$$

For deriving the solution of the problem (8.26), the LINGO-18 codes are:

> **LINGO-18 codes for the problem (8.26)**
> MODEL:
> MAX= A1 + A2;
> 84 * A1 - 5 * X1 + 4 * X2 <= 24;
> 82 * A2 + 2 * X1 - 8 * X2 <= 22;
> - X1 + X2 <= 6;
> X1 <= 12;
> X1 + X2 <= 16;
> X2 <= 8;
> X1 >= 0;
> X2 >= 0;
> END

Solving the FGP will give $\mu_1 = 0.3809524$, $\mu_2 = 0.8536585, x_1 = x_2 = 8$. This implies that goal 1 will be achieved by 38% and goal 2 by 85% respectively.

8.4 Multi-Objective Optimization Problem with Fuzzy Numbers

In the linear optimization problem, fuzzy set theory was first introduced by [39]. In a LPP he considered a fuzzy goal and constraints. He proved that an equivalent LPP exists, by following the fuzzy decision approach proposed by [1] along with linear membership functions. In any optimization problem either single, multiple, linear or non-linear, fuzziness can occur. The elements of either objective function or constraint or both can be expressed as fuzzy numbers rather than crisp numbers. Ultimately, the solution of the fuzzy problem can either be a fuzzy set or crisp solution.

Consider a MOOP as follows:

$$\left. \begin{array}{c} \min \text{ (or max)} \begin{pmatrix} z_1(\underline{x}) \\ z_2(\underline{x}) \\ \vdots \\ z_k(\underline{x}) \end{pmatrix} \\ \text{subject to } g_i(\underline{x})\{\leq, =, \geq\} \ b_i, \ i = 1, 2, \ldots, m \\ \underline{x} \geq 0 \end{array} \right\} \quad (8.27)$$

where $z_k(\underline{x})$; $k = 1, 2, \ldots, K$ is the kth objective function.

Suppose for each objective function DM has fuzzy coefficients, then to solve the MOOP (8.27) follow the algorithm discussed in the next section.

8.4.1 Algorithm

Step 1 If any parameter is in terms of fuzzy numbers, convert it into crisp by any defuzzification method first. Then, solve the stated problem as a single objective problem using only one objective at a time and ignoring the other objective functions. The solutions thus obtained are considered to be ideal solutions for the objectives.

Step 2 The ideal solutions of each objective function are then used to calculate the value of all other objectives. Find the goal level for each objective function by calculating the maximum and a minimum value of the kth objective function. Then, obtain the upper and lower tolerance limits for each objective function, i.e., $U_k = \max(z_k)$ and $L_k = \min(z_k)$; $k = 1, 2, \ldots, K$.

Step 3 Set up the linear membership function for the given objective functions as:
When the objective function is of minimization type:

$$\omega_j(z_j(\underline{x})) = \begin{cases} 0 & \text{if } z_j(\underline{x}) \geq L_j \\ \frac{z_j - L_j}{U_j - L_j} & \text{if } L_j \leq z_j(\underline{x}) \leq U_j \quad ; \quad j = 1, 2, \ldots, k \\ 1 & \text{if } z_j(\underline{x}) \leq U_j \end{cases}$$

When the objective function is of maximization type:

$$\omega_j(z_j(\underline{x})) = \begin{cases} 0 & \text{if } z_j(\underline{x}) \leq L_j \\ \frac{U_j - z_j}{U_j - L_j} & \text{if } L_j \leq z_j(\underline{x}) \leq U_j \quad ; \quad j = 1, 2, \ldots, k \\ 1 & \text{if } z_j(\underline{x}) \geq U_j \end{cases}$$

Step 4 Construct the fuzzy programming model using the membership function and following [1] as:

$$\left. \begin{array}{l} \max \lambda \\ \text{subject to } \lambda \leq \omega_j(z_j(\underline{x})), \quad j = 1, 2, \ldots, k \\ \quad\quad\quad\quad g_i(\underline{x})\{\leq, =, \geq\} \; b_i, \quad i = 1, 2, \ldots, m \\ \quad\quad\quad\quad \text{and } \underline{x} \geq 0 \end{array} \right\} \quad (8.28)$$

Now, solve it using optimization software LINGO-18. After solving the model, we get the most preferred fuzzy efficient, and compromised solution for MOOP.

8.5 Numerical Example using LINGO-18

For demonstration purpose, hypothetical numerical data has been taken. Consider a manufacturing company manager produces two types of products A and B that are processed on two different machines X & Y. In view of a real-life

Table 8.3: Data for Different Parameters.

Products	Processing Time on Machine		Cost	Revenue	Demand
	X	Y			
A	(1,3,5)	(2,4,5)	(28, 30, 31)	(18, 8, 7)	(2, 6, 8)
B	(2,4,6)	(1,2,3)	(42, 45, 46)	(13, 11, 4)	(6, 8, 10)
Availability	(7,8,10)	(5,7,9)			

problem, all the parameters are considered triangular fuzzy numbers and the data about the parameters are given below in Table (8.3).

Now, to optimize the revenue and cost of the company, we determine how many batches of product A and product B have to be made. For that, we have formulated the problem into a fuzzy multi-objective programming problem. Using the data given in Table (8.3), the fuzzy multi-objective programming problem will be:

$$\left. \begin{array}{l} \max \; \tilde{R} = (18,8,7)x_1 + (13,11,4)x_2 \\ \min \; \tilde{C} = (28,30,31)x_1 + (42,45,46)x_2 \\ \text{subject to } (1,3,5)x_1 + (2,4,5)x_2 \leq (7,8,10) \\ \qquad (2,4,6)x_1 + (1,2,3)x_2 \leq (5,7,9) \\ \qquad (2,4,5)x_1 \geq (2,6,8) \\ \qquad (1,3,5)x_2 \leq (6,8,10) \\ \qquad x_i \geq 0; \; i = 1,2 \end{array} \right\} \qquad (8.29)$$

Firstly, all the parameters represented by triangular fuzzy numbers are converted into crisp numbers by α-cut defuzzification method as discussed in Chapter 4. Then, following the steps of the algorithm, and solving the MOOP by taking one objective at a time, the individual optimum solution and the upper & lower tolerance limits are computed as:

LINGO-18 codes for revenue objective function:

MODEL:
MAX= (7 + A) * X1 + (4 + 7 * A) * X2;
(2 * A + 1) * X1 + (2 * A + 2) * X2 <= (A + 7);
(2 * A + 2) * X1 + (A + 1) * X2 <= (2 * A + 5);
(5 - A) * X1 >= (8 - 2 * A);
(2 * A + 1) * X2 <= (2 * A + 6);
A = 0.5;
END

LINGO-18 codes for cost objective function:
MODEL:
MIN= (2 * A + 28) * X1 + (3 * A + 42) * X2;
(2 * A + 1) * X1 + (2 * A + 2) * X2 <= (A + 7);
(2 * A + 2) * X1 + (A + 1) * X2 <= (2 * A + 5);
(5 - A) * X1 >= (8 - 2 * A);
(2 * A + 1) * X2 <= (2 * A + 6);
A = 0.5;
END

The results are as follows:

$x_1 = 1.56, x_2 = 0.89$ with $R = 18.33$ and $x_1 = 1.56, x_2 = 0$ with $C = 45.11$

$$R^U = 18.33, R^L = 11.70, C^U = 83.96, \ \& \ C^L = 45.11$$

Now, the fuzzy model with linear membership function will be given as:

max λ

subject to $\lambda \leq \dfrac{(((7+\alpha)x_1 + (4+7\alpha)x_2) - 11.70)}{6.63}$

$\lambda \leq \dfrac{(83.96 - ((2\alpha+28)x_1 + (3\alpha+42)x_2))}{38.85}$

$(2\alpha + 1)x_1 + (2\alpha + 2)x_2 \leq (\alpha + 7)$ (8.30)

$(2\alpha + 2)x_1 + (\alpha + 1)x_2 \leq (2\alpha + 5)$

$(5 - \alpha)x_1 \geq (8 - 2\alpha)$

$(2\alpha + 1)x_2 \leq (2\alpha + 6)$

$x_i \geq 0; \ i = 1, 2$

The problem (8.30) is solved by an optimization software LINGO-18 and the compromise solution is derived as follows:

LINGO-18 codes for fuzzy model:
MODEL:
MAX= C;
C <= (((7 + A) * X1 + (4 + 7 * A) * X2) - 11.7) / 6.63;
C <= (83.96 - ((2 * A + 28) * X1 + (3 * A + 42) * X2)) / 38.85;
(2 * A + 1) * X1 + (2 * A + 2) * X2 <= (A + 7);
(2 * A + 2) * X1 + (A + 1) * X2 <= (2 * A + 5);
(5 - A) * X1 >= (8 - 2 * A);
(2 * A + 1) * X2 <= (2 * A + 6);
A = 0.5;
END

$x_1 = 1.94, x_2 = 0.13$ with $R = 15.53 \approx \$16$ and $C = 61.92 \approx \$62$

Chapter 9

Applications of Optimization

Optimize means maximize or minimize or, in other words; optimization means to obtain the best out of the available resources. In every sphere of life, everyone wants to optimize their resources, decisions, and each activity of life. Therefore, in every field optimization is applicable. In this chapter applicability of optimization has been shown in some of the business areas such as finance, marketing, media, supply chain, human resource, and others.

9.1 Optimization Problems in Finance

Finance is an essential part of every individual as well as business. If the finances are not managed properly, it will create problems in any business or individual's life. Optimization can be used in financial decision-making that involves capital budgeting, make-or-buy, asset allocation, portfolio selection, financial planning, and more. In Portfolio selection problems, decision-making involves choosing specific investments such as stocks, mutual funds, fixed deposits, bonds, and many more investment alternatives. Such types of problems are faced by portfolio managers of banks, mutual funds, and insurance companies. Managers or investors need to take optimal decisions to achieve the objectives of either expected return maximization or minimization of risk portfolio.

Example 120 *A person is interested in investing $6,546 in a mix of investments. The investment choices & expected rates of return on each one of them are represented in Table (9.1).*

Table 9.1: Expected Returns of the Investment Alternatives.

Investment	Projected Rate of Return
Mutual Fund A	0.12
Mutual Fund B	0.09
Money Market Fund	0.08
Government Bonds	0.085
Share X	0.18
Share Y	0.16

The investor has the following conditions for investments:

- He wants at least 35% of his investment in government bonds.

- He has specified that the combined investment in two shares (Because of the higher perceived risk of the two shares) should not exceed $1,047.

- At least 20% of the investment should be in the money market fund.

- The amount of money invested in shares should not exceed the amount invested in mutual funds.

- Finally, he wants that the amount invested in mutual fund A should be more than the amount invested in mutual fund B.

Formulate as LP problem to get highest total return.

Solution   Decision Variables:
x_1 - amount of investment in Mutual Fund A
x_2 - amount of investment in Mutual Fund B
x_3 - amount of investment in Money Market Fund
x_4 - amount of investment in Government Bonds
x_5 - amount of investment in Share X
x_6 - amount of investment in Share Y

Objective Function: The objective is to maximize the total return on investments, i.e.,

$$\max R = 0.12x_1 + 0.09x_2 + 0.08x_3 + 0.085x_4 + 0.18x_5 + 0.16x_6$$

Constraint 1: Total Investment amount

$$x_1 + x_2 + x_3 + x_4 + x_5 + x_6 = 6,546$$

Constraint 2: Investments in government bonds

$$x_4 \geq 2,291 (35\% \text{ of } 6,549)$$

Constraint 3: Investments in two shares

$$x_5 + x_6 \leq 1,0470$$

Constraint 4: Investments in money market funds

$$x_3 \geq 1,309 (20\% \text{ of } 6,549)$$

Constraint 5: Proportion of investments in shares and mutual funds

$$x_5 + x_6 \leq x_1 + x_2$$

Constraint 6:
$$x_1 \geq x_2$$

Non-Negativity Restriction:

$$x_1, x_2, x_3, x_4, x_5, x_6 \geq 0$$

Therefore, the LP model of the problem is as follows:

$$\left.\begin{aligned}
\max \ R &= 0.12x_1 + 0.09x_2 + 0.08x_3 + 0.085x_4 + 0.18x_5 + 0.16x_6 \\
\text{subject to } & x_1 + x_2 + x_3 + x_4 + x_5 + x_6 = 6,546 \\
& x_4 \geq 2,291 \\
& x_5 + x_6 \leq 1,047 \\
& x_3 \geq 1,309 \\
& x_5 + x_6 \leq x_1 + x_2 \\
& x_1 \geq x_2 \\
\text{and } & x_1, x_2, x_3, x_4, x_5, x_6 \geq 0
\end{aligned}\right\} \quad (9.1)$$

LINGO-18 codes for problem (9.1)

$MODEL$:
$SETS$:
$product/1..6/ : R, x;$
$const/1..3/ : mat;$
$const1/1/ : a;$
$const2/1..2/ : mat1;$
$material(const, product) : b;$
$material1(const1, product) : s;$
$material2(const2, product) : d;$
$endsets$
!Objective function of maximization of average return;
$[objective]max = @sum(product(j) : R(j) * x(j));$
!Constraints;
$@for(const(i) : [Material\_constraints]$
$@sum(product(j) : b(i,j) * x(j)) >= mat(i));$
$@for(const1(i) : [Material1\_constraints]$
$@sum(product(j) : s(i,j) * x(j)) = a(i));$
$@for(const2(i) : [Material2\_constraints]$
$@sum(product(j) : d(i,j) * x(j)) <= mat1(i));$
$Data$:
$R = 0.12, 0.09, 0.08, 0.085, 0.18, 0.16;$
$mat = 2291, 1309, 0;$
$a = 6546;$
$mat1 = 1047, 0;$
$b = 0\ 0\ 0\ 1\ 0\ 0\quad 0\ 0\ 1\ 0\ 0\ 0\quad 1\ -1\ 0\ 0\ 0\ 0;$
$s = 1\ 1\ 1\ 1\ 1\ 1;$
$d = 0\ 0\ 0\ 0\ 1\ 1\quad -1\ -1\ 0\ 0\ 1\ 1;$
$enddata$
end

The optimal solution of LPP (9.1) derived through LINGO-18 software is $x_1 = 1,899$, $x_2 = 0$, $x_3 = 1,309$, $x_4 = 2,291$, $x_5 = 1,047$, $x_6 = 0$ with maximum total return $715.80. ∎

Example 121 *A company ABC wants to invest his shares in ten prominent mutual funds of Large cap category. The expected average percentage of return and risk of ten mutual funds are given in Table (9.2).*

CEO of the company does not want to invest more than 40% on Reliance Large Cap Fund. Also, he wants the average risk for the portfolio should not exceed 5. The number of investments in the Edelweiss Large Cap Fund must be higher than the HDFC Top 100 Fund investment. Formulate the LP model to maximize the average return of the company.

Table 9.2: Average Return and Risk Percentage of the Investment Alternatives.

S. No.	Mutual Funds	Average Return (%)	Total Risk (SD)%
1	Canara Robeco Blue Chip Equity	0.86	3.88
2	Edelweiss Large Cap Fund	0.95	3.86
3	HDFC Top 100 Fund	1.03	4.61
4	Axis Bluechip Fund - Growth	1.00	3.85
5	Invesco India Largecap Fund	0.95	3.72
6	UTI Master Share-Growth	0.81	3.64
7	ICICI Prudential Bluechip Fund	1.05	3.72
8	Reliance Large Cap Fund	1.18	4.32
9	IDFC Large Cap Fund	0.81	3.64
10	JM Core 11 Fund	1.07	5.77

Solution   Decision Variables:

x_1- proportion of investment in Canara Robeco Blue Chip Equity
x_2- proportion of investment in Edelweiss Large Cap Fund
x_3- proportion of investment in HDFC Top 100 Fund
x_4- proportion of investment in Axis Bluechip Fund - Growth
x_5- proportion of investment in Invesco India Largecap Fund
x_6- proportion of investment in UTI Master Share-Growth
x_7- proportion of investment in ICICI Prudential Bluechip Fund
x_8- proportion of investment in Reliance Large Cap Fund
x_9- proportion of investment in IDFC Large Cap Fund
x_{10}- proportion of investment in JM Core 11 Fund

Objective Function: The objective is to maximize the total return on investments, i.e.,

$$\max\ R = 0.86x_1 + 0.95x_2 + 1.03x_3 + x_4 + 0.95x_5 + 0.81x_6 + 1.05x_7 + 1.18x_8 + 0.81x_9 + 1.07x_{10}$$

Constraint 1:

$$x_8 \leq 0.40$$

Constraint 2:

$$3.88x_1 + 3.86x_2 + 4.61x_3 + 3.85x_4 + 3.72x_5 + 3.64x_6 + 3.72x_7 + 4.32x_8 + 3.64x_9 + 5.77x_{10} \leq 5$$

Constraint 3:

$$x_2 \geq x_3$$

Constraint 4: Investments proportions

$$x_1 + x_2 + x_3 + x_4 + x_5 + x_6 + x_7 + x_8 + x_9 + x_{10} = 1$$

Non-Negativity Restriction:

$$x_1, x_2, x_3, x_4, x_5, x_6, x_7, x_8, x_9, x_{10} \geq 0$$

Therefore, the LP model of the problem is as follows:

$$\left.\begin{aligned}
\max\ R &= 0.86x_1 + 0.95x_2 + 1.03x_3 + x_4 + 0.95x_5 \\
&\quad + 0.81x_6 + 1.05x_7 + 1.18x_8 + 0.81x_9 + 1.07x_{10} \\
\text{subject to}\ & x_8 \leq 0.40 \\
& 3.88x_1 + 3.86x_2 + 4.61x_3 + 3.85x_4 + 3.72x_5 \\
&\quad + 3.64x_6 + 3.72x_7 + 4.32x_8 + 3.64x_9 + 5.77x_{10} \leq 5 \\
& x_2 \geq x_3 \\
& x_1 + x_2 + x_3 + x_4 + x_5 + x_6 + x_7 + x_8 + x_9 + x_{10} = 1 \\
\text{and}\ & x_1, x_2, x_3, x_4, x_5, x_6, x_7, x_8, x_9, x_{10} \geq 0
\end{aligned}\right\} \quad (9.2)$$

LINGO-18 codes for problem (9.2)

$MODEL:$
$SETS:$
$product/1..10/ : R, x;$
$const/1..2/ : mat;$
$const1/1/ : a;$
$const2/1/ : c;$
$material(const, product) : b;$
$material1(const1, product) : s;$
$material2(const2, product) : d;$
$endsets$
!Objective function of maximization of average return;
$[objective]max = @sum(product(j) : R(j)*x(j));$
!Constraints;
$@for(const(i) : [Material\_constraints]$
$@sum(product(j) : b(i,j)*x(j)) <= mat(i));$
$@for(const1(i) : [Material1\_constraints]$
$@sum(product(j) : s(i,j)*x(j)) >= a(i));$
$@for(const2(i) : [Material2\_constraints]$
$@sum(product(j) : d(i,j)*x(j)) = c(i));$
$Data:$
$R = 0.86, 0.95, 1.03, 1, 0.95, 0.81, 1.05, 1.18, 0.81, 1.07;$
$mat = 0.40, 5;$
$a = 0;$
$c = 1;$
$b = 0\ 0\ 0\ 0\ 0\ 0\ 1\ 0\ 0\quad 3.88\ 3.86\ 4.61\ 3.85\ 3.72\ 3.64\ 3.72\ 4.32\ 3.64\ 5.77;$
$s = 0\ 1\ -1\ 0\ 0\ 0\ 0\ 0\ 0\ 0;$
$d = 1\ 1\ 1\ 1\ 1\ 1\ 1\ 1\ 1\ 1;$
$enddata$
end

The optimal solution of LPP (9.2) derived through LINGO-18 software is $x_1 = 0$, $x_2 = 0$, $x_3 = 0$, $x_4 = 0$, $x_5 = 0$, $x_6 = 0$, $x_7 = 0.10$, $x_8 = 0.40$, $x_9 = 0$, $x_{10} = 0.50$ with maximum total return 1.112. ∎

9.2 Optimization Problems in Marketing

In marketing, constraint optimization is useful in media selection. Marketing managers can get help from a linear programming technique allocating a fixed budget to various advertising media. The problem can consider the objectives such as maximizing reach, frequency, and quality of exposure under the allowable allocation restrictions usually arise during consideration of company policy, contract requirements, and media availability. In a marketing application, optimization can also be used to minimize the cost of marketing with the fulfilment of client's needs, maximization of sales with some of the company restrictions such as budget, and number of advertisements.

Example 122 *A chocolate production company wishes to plan an advertising campaign for its product in four different media, viz., television, radio, and magazine. The primary aim of advertising is to reach the maximum number of customers. From a market survey, the following data is obtained:*

Table 9.3: Market Survey Data.

	Television		Radio	Magazine
	Prime Day	Prime Time		
Cost of an advertising Unit	$556	$1,043	$417	$209
Number of potential customers reached per unit	4,00,000	9,00,000	5,00,000	2,00,000
Number of women customers reached per unit	3,00,000	4,00,000	2,00,000	1,00,000

The company has a budget of only $11,129 for advertising expenditure. Further it is required that

- *Among women at least 2 million exposures take place.*

- *Television advertising cost should be limited to $6,955.*

- *At least 5 advertising units be bought on television (combining prime day and prime time).*

- *The number of advertising units on radio and magazine should range between 5 and 12.*

To maximize the potential reach, formulate the problem as a LP model.

Solution **Decision Variables:**
x_1- Number of advertising units bought in prime day on television
x_2- Number of advertising units bought in prime time on television
x_3- Number of advertising units bought in radio
x_4- Number of advertising units bought in magazine

Objective Function:

$$\max Z = 4,00,000x_1 + 9,00,000x_2 + 5,00,000x_3 + 2,00,000x_4$$

Constraint 1: Total Advertising Budget

$$556x_1 + 1,043x_2 + 417x_3 + 209x_4 \leq 11,129$$

Constraint 2: Women Customers Reach Count

$$3,00,000x_1 + 4,00,000x_2 + 2,00,000x_3 + 1,00,000x_4 \geq 20,00,000$$

$$\text{or} \quad 3x_1 + 4x_2 + 2x_3 + x_4 \geq 20$$

Constraint 3: Television budget

$$556x_1 + 1,043x_2 \leq 6,995$$

Constraint 4: Number of advertising units on television

$$x_1 + x_2 \geq 5$$

Constraint 5: Limits of advertisements on radio

$$5 \leq x_3 \leq 12$$

Constraint 6: Limits of advertisements on magazines

$$5 \leq x_4 \leq 12$$

Non-Negativity Restriction:

$$x_1, x_2, x_3, x_4 \geq 0$$

Therefore, the LP model of the problem is as follows:

$$\left.\begin{aligned}
\max \ Z &= 4,00,000x_1 + 9,00,000x_2 + 5,00,000x_3 + 2,00,000x_4 \\
\text{subject to} \ \ & 556x_1 + 1,043x_2 + 417x_3 + 209x_4 \leq 11,129 \\
& 3x_1 + 4x_2 + 2x_3 + x_4 \geq 20 \\
& 556x_1 + 1,043x_2 \leq 6,995 \\
& x_1 + x_2 \geq 5 \\
& 5 \leq x_3 \leq 12 \\
& 5 \leq x_4 \leq 12 \\
\text{and} \ \ & x_1, x_2, x_3, x_4 \geq 0
\end{aligned}\right\} \quad (9.3)$$

> **LINGO-18 codes for problem (9.3)**
> $MODEL:$
> $SETS:$
> $product/1..4/:C,x;$
> $const/1..4/:mat;$
> $const1/1..4/:mat1;$
> $material(const, product):a;$
> $material1(const1, product):b;$
> $endsets$
> $[objective]max = @sum(product(j):C(j)*x(j));$
> $@for(const(i):[Material\_constraints]$
> $@sum(product(j):a(i,j)*x(j)) <= mat(i));$
> $@for(const1(i):[Material1\_constraints]$
> $@sum(product(j):b(i,j)*x(j)) >= mat1(i));$
> $Data:$
> $C = 4,9,5,2;$
> $mat = 11129, 6995, 12, 12;$
> $mat1 = 20, 5, 5, 5;$
> $a = 556\ 1043\ 417\ 209\ \ 556\ 1043\ 0\ 0\ \ 0\ 0\ 1\ 0\ \ 0\ 0\ 0\ 1;$
> $b = 3\ 4\ 2\ 1\ \ 1\ 1\ 0\ 0\ \ 0\ 0\ 1\ 0\ \ 0\ 0\ 0\ 1;$
> $enddata$
> end

The optimal solution of LPP (9.3) derived through LINGO-18 software is $x_1 = 0.277$, $x_2 = 4.723$, $x_3 = 12$, $x_4 = 5$ with maximum reach 113.61. ∎

Example 123 *Suppose chief marketing officer of a company XYZ tasked the company's interns to suggest allocation of budgets, which should not exceed $5 million across TV, Print (catalogue, billboards, Airports), SEO, AdWords, Facebook, and Mobile. Also, there are about 2.5 million customers in the target market segment. From the survey, ROI from each campaign and customer reach has been determined as given in Table (9.4).*

Table 9.4: ROI and Customer Reach Data.

Media	ROI (%)	Customer Reach
TV	7	2
Print	3	0.3
Mobile	15	1.8
SEO+AdWords	12	0.9
Facebook	5	2

From the Multi-Channel Funnel Analysis and experience, the following marketing strategy has been adopted.

- Production and airing the TV advertisements costs at least $700K.
- Print media should not account for more than 5% of the budget.
- Minimum Ad agency costs for content creation and placement for Billboards, game place promotions, airport promotion is $400K.
- Mobile and SEO + AdWords together should account for at least 50%, and SEO + AdWords should be no more than 2.5 times the mobile marketing costs.
- Mobile content and advertising cost anywhere between $500K and $1100K.
- Agency costs for Facebook marketing cost at least $100K.

Obtain the optimal budget allocation to maximize the ROI by using the Linear Programming technique.

Solution   Decision Variables:
x_1- Budget for TV
x_2- Budget for Print Media
x_3- Budget for Mobile
x_4- Budget for SEO+AdWords
x_5- Budget for Facebook Marketing

Objective Function:

$$\max ROI = 0.07x_1 + 0.03x_2 + 0.15x_3 + 0.12x_4 + 0.05x_5$$

Constraint 1: Total Budget

$$x_1 + x_2 + x_3 + x_4 + x_5 \leq 5000$$

Constraint 2: TV advertising costs

$$x_1 \geq 700$$

Constraint 3: Print Media budget

$$x_2 \leq 0.05(x_1 + x_2 + x_3 + x_4 + x_5)$$

Constraint 4: Print Media cost

$$x_2 \geq 400$$

Constraint 5: Mobile and SEO+AdWords budget

$$x_3 + x_4 \geq 0.5(x_1 + x_2 + x_3 + x_4 + x_5)$$

Constraint 6:

$$x_4 \leq 2.5x_3$$

Constraint 7: Mobile Marketing budget

$$500 \leq x_4 \leq 1100$$

Constraint 8: Facebook Advertising Cost

$$x_5 \geq 100$$

Constraint 9: Reach

$$2x_1 + 0.3x_2 + 1.8x_3 + 0.9x_4 + 2x_5 \leq 2500$$

Non-Negativity Restriction:

$$x_1, x_2, x_3, x_4 \geq 0$$

Therefore, the LP model of the problem is as follows:

$$\left.\begin{aligned}
\max \ \text{ROI} &= 0.07x_1 + 0.03x_2 + 0.15x_3 + 0.12x_4 + 0.05x_5 \\
\text{subject to } & x_1 + x_2 + x_3 + x_4 + x_5 \leq 5000 \\
& x_1 \geq 700 \\
& x_2 \leq 0.05(x_1 + x_2 + x_3 + x_4 + x_5) \\
& x_2 \geq 400 \\
& x_3 + x_4 \geq 0.5(x_1 + x_2 + x_3 + x_4 + x_5) \\
& x_4 \leq 2.5x_3 \\
& 500 \leq x_4 \leq 1100 \\
& x_5 \geq 100 \\
& 2x_1 + 0.3x_2 + 1.8x_3 + 0.9x_4 + 2x_5 \leq 2500 \\
\text{and } & x_1, x_2, x_3, x_4, x_5 \geq 0
\end{aligned}\right\} \quad (9.4)$$

LINGO-18 codes for problem (9.4)

$MODEL:$
$SETS:$
$product/1..5/: C, x;$
$const1/1..2/: mat;$
$const2/1..3/: mat1;$
$const3/1/::;$
$const4/1/::;$
$const5/1/: r;$
$material(const1, product): b;$
$material1(const2, product): s;$
$material2(const3, product): f;$
$material3(const4, product): g;$
$material4(const5, product): e;$
$endsets$
$[objective]max = @sum(product(j): C(j)*x(j));$
$@for(const1(i): [Material\_constraints]$
$@sum(product(j): b(i,j)*x(j)) <= mat(i));$
$@for(const2(i): [Material1\_constraints]$
$@sum(product(j): s(i,j)*x(j)) >= mat1(i));$
$@for(const3(i): [Material2\_constraints]$
$@sum(product(j): f(i,j)*x(j)) <= @sum(product(j): 0.05*x(j)));$
$@for(const4(i): [Material3\_constraints]$
$@sum(product(j): g(i,j)*x(j)) >= @sum(product(j): 0.5*x(j)));$
$@for(const5(i): [Material4\_constraints]$
$@sum(product(j): e(i,j)*x(j)) <= @sum(product(j): 2.5*r(i,j)));$
$Data:$
$C = 0.07, 0.03, 0.15, 0.12, 0.05;$
$mat = 5000, 2500;$
$mat1 = 700, 400, 100;$
$b = 1\ 1\ 1\ 1\ 1\ \ 2\ 0.3\ 1.8\ 0.9\ 2;$
$s = 1\ 0\ 0\ 0\ 0\ \ 0\ 1\ 0\ 0\ 0\ \ 0\ 0\ 0\ 0\ 1;$
$f = 0\ 1\ 0\ 0\ 0;$
$g = 0\ 0\ 1\ 1\ 0;$
$r = 0\ 0\ 1\ 0\ 0;$
$enddata$
end

The optimal solution of LPP (9.4) derived through LINGO-18 software is $x_1 = 700$, $x_2 = 400$, $x_3 = 268.9$, $x_4 = 551.2$, $x_5 = 0$. The current solution is infeasible which shows that problem can be reformulated or any method to remove the infeasibility can be used. ∎

9.3 Optimization Problems in Human Resource Management

In today's competitive world, every organization, be it industry for production, a unit of manufacturing, a government organization, a limited corporation, a banking sector, or an educational institution, is focusing on increasing profit margin. And to do that, the most important factor is achieving maximum output with minimum resources (input). Therefore, it has become essential for organizations to plan the use of resources in the most optimum way. Resources can be anything that acts as an input. They can be men, material, money, or other assets that can be used by an organization or an individual to function effectively. Linear programming is one of the widely used optimization techniques to optimize the scare resources. Therefore, it can be applied in managing human resources as well.

Example 124 *A university XYZ wants to optimally allocate the courses to the teachers according to their abilities. After appropriate introspection and evaluation of each teacher's ability to teach a particular course, an efficiency rating of 100 has been allotted. The evaluation criteria include past results, student feedbacks, subject knowledge, research on the area, past developments, etc. Table (9.5) below shows the relative ratings of a teacher to a course.*

Table 9.5: Relative Ratings.

Faculty	Courses				
	Mathematics	Marketing	Finance	Operations	RM
A	70	76	88	87	84
B	79	80	82	90	89
C	82	84	77	81	86
D	88	89	85	81	77
E	92	74	82	84	86

The problem is how the faculties should be assigned to the courses to maximize the educational quality in the department. As we know that the assignment problems usually deal with the minimization situations, and the calculations are also accessible in that way, so the above maximization problem is reduced to a minimization problem by finding the regrets matrix, as shown in Table (9.6).

Solution Decision Variables:

$$x_{ij} = \begin{cases} 1, & \text{if } i^{th} \text{ faculty is assigned to } j^{th} \text{ course} \\ 0, & \text{otherwise} \end{cases}$$

Table 9.6: Regrets Matrix.

Faculty	Courses				
	Mathematics	Marketing	Finance	Operations	RM
A	30	24	12	13	16
B	21	20	18	10	11
C	18	16	23	19	14
D	12	11	15	19	23
E	08	26	18	16	14

Objective Function:

$$\min Z = \sum_{i=1}^{n}\sum_{j=1}^{n} R_{ij} x_{ij}$$

where R_{ij} denotes the regret value given in Table (9.6).

Constraint 1: Row Constraints

$$\sum_{j=1}^{n} x_{ij} = 1$$

Constraint 2: Column Constraints

$$\sum_{i=1}^{n} x_{ij} = 1$$

Binary Restriction:

$$x_{ij} = 0/1$$

Therefore, the LP model of the problem is as follows:

$$\begin{aligned}
\min\ Z = & 30x_{11} + 24x_{12} + 12x_{13} + 13x_{14} + 16x_{15} + 21x_{21} + 20x_{22} + 18x_{23} \\
& + 10x_{24} + 11x_{25} + 18x_{31} + 16x_{32} + 23x_{33} + 19x_{34} + 14x_{35} + 12x_{41} \\
& + 11x_{42} + 15x_{43} + 19x_{44} + 23x_{45} + 8x_{51} + 26x_{52} + 18x_{53} + 18x_{54} + 14x_{55} \\
\text{subject to}\ & \sum_{i=1}^{n} x_{ij} = 1 \\
& \sum_{j=1}^{n} x_{ij} = 1 \\
\text{and}\ & x_{ij} = 0/1
\end{aligned}$$

(9.5)

> **LINGO-18 codes for problem (9.5)**
>
> $MODEL:$
> !Faculty Assignment Problem;
> $SETS:$
> $Faculty/A\ B\ C\ D\ E/;$
> $Courses/Mth\ Mark\ Fin\ Oper\ RM/;$
> $Links(Faculty, Courses): Cost, Assign;$
> $endsets$
> !Objective;
> $min = @sum(Links(i,j): Cost(i,j) * Assign(i,j));$
> !Constraints;
> $@for(Faculty(i): @sum(Courses(j): Assign(i,j)) = 1);$
> $@for(Courses(j): @sum(Faculty(i): Assign(i,j)) = 1);$
> $@for(Links(i,j): @BIN(Assign(i,j)));$
> $Data:$
> $Cost = 30\ 24\ 12\ 13\ 16$
> $21\ 20\ 18\ 10\ 11$
> $18\ 16\ 23\ 19\ 14$
> $12\ 11\ 15\ 19\ 23$
> $8\ 26\ 18\ 16\ 14;$
> $enddata$
> end

The optimal solution of LPP (9.5) derived through LINGO-18 software is $x_{13} = 1$, $x_{24} = 1$, $x_{35} = 1$, $x_{42} = 1$, $x_{51} = 1$ with minimum Z = 55. ∎

9.4 Vendor Selection Optimization Problem

In today's highly competitive and dynamic business scenario, the vendor/supplier selection is a crucial problem involving many qualitative and quantitative criteria. Suppliers are considered key to a firm's ability to provide quality products in a shorter time at lower costs and with greater flexibility and reduced risk.

Example 125 *Mr Rahul is working as an operations analyst in a company. His manager wants him to solve the problem of selecting the best vendors for the company's products. The company has four vendors and manufactures four types of products. The problem is to maximize the preference weights for the vendor's selection. The weighted values are given in Table (9.7).*
Manager also had some conditions:

- *For every product at least one vendor should be selected.*

- *Maximum 3, 2, 2, 2 products should be allocated to 4 vendors.*

- *Total number of products should not be more than 9.*

Table 9.7: Preference Weights.

x_{ij}; i – vendor index; j – product index	Weights (%)
$(x_{11}, x_{12}, x_{13}, x_{14})$	(0.334, 0.47, 0.112, 0.219)
$(x_{21}, x_{22}, x_{23}, x_{24})$	(0.332, 0.233, 0.621, 0.179)
$(x_{31}, x_{32}, x_{33}, x_{34})$	(0.401, 0.433, 0.116, 0.579)
$(x_{41}, x_{42}, x_{43}, x_{44})$	(0.279, 0.511, 0.62, 0.222)

Formulate the following problem as an LP model.

Solution   **Decision Variables:**

$$x_{ij} = \begin{cases} 1, & \text{if } i^{th} \text{ vendor is selected for } j^{th} \text{ product} \\ 0, & \text{otherwise} \end{cases}$$

Objective Function:

$$\max Z = \sum_{i=1}^{n}\sum_{j=1}^{n} w_{ij} x_{ij}$$

where w_{ij} denotes the preference weights given in Table 9.7.

Constraint 1: Vendor limitation

$$\sum_{i=1}^{n} x_{ij} \geq 1$$

Constraint 2: Limits of products that a vendor can provide

$$x_{11} + x_{12} + x_{13} + x_{14} \leq 3$$

$$x_{21} + x_{22} + x_{23} + x_{24} \leq 2$$

$$x_{31} + x_{32} + x_{33} + x_{34} \leq 2$$

$$x_{41} + x_{42} + x_{43} + x_{44} \leq 2$$

Constraint 3: Maximum product limitation

$$x_{11} + x_{12} + x_{13} + x_{14} + x_{21} + x_{22} + x_{23} + x_{24} + x_{31} + x_{32} + x_{33} + x_{34} + x_{41} + x_{42} + x_{43} + x_{44} \leq 9$$

Binary Restriction:

$$x_{ij} = 0/1$$

Therefore, the LP model of the problem is as follows:

$$\left.\begin{aligned}
\max \ Z = & 0.334x_{11} + 0.47x_{12} + 0.112x_{13} + 0.219x_{14} + 0.332x_{21} + 0.233x_{22} + \\
& 0.621x_{23} + 0.179x_{24} + 0.401x_{31} + 0.433x_{32} + 0.116x_{33} + 0.579x_{34} \\
& + 0.279x_{41} + 0.511x_{42} + 0.62x_{43} + 0.222x_{44} \\
\text{subject to} \ & \sum_{i=1}^{n} x_{ij} = 1 \\
& \sum_{i=1}^{n} x_{ij} = 1 \\
& x_{11} + x_{12} + x_{13} + x_{14} \leq 3 \\
& x_{21} + x_{22} + x_{23} + x_{24} \leq 2 \\
& x_{31} + x_{32} + x_{33} + x_{34} \leq 2 \\
& x_{41} + x_{42} + x_{43} + x_{44} \leq 2 \\
& x_{11} + x_{12} + x_{13} + x_{14} + x_{21} + x_{22} + x_{23} + x_{24} + x_{31} + x_{32} + x_{33} \\
& + x_{34} + x_{41} + x_{42} + x_{43} + x_{44} \leq 9 \\
\text{and} \ & x_{ij} = 0/1
\end{aligned}\right\} \tag{9.6}$$

> **LINGO-18 codes for problem** (9.6)
> $MODEL:$
> $SETS:$
> $Vendor:V;$
> $Product:P,M;$
> $Links(Vendor, Product):W,X;$
> $endsets$
> $!Objective;$
> $min = @sum(Links(i,j):W(i,j)*X(i,j));$
> $!Constraints;$
> $@for(Product(j):@sum(Vendor(i):X(i,j))=1);$
> $@for(Vendor(i):@sum(Product(j):X(i,j))=1);$
> $@for(Vendor(i):@sum(Product(j):X(i,j))<=M(i));$
> $@sum(Links(i,j):X(i,j))<=A;$
> $@for(Links(i,j):@BIN(X(i,j)));$
> $Data:$
> $Vendor = V1\ V2\ V3\ V4;$
> $Product = P1\ P2\ P3\ P4;$
> $W = 0.334\ 0.47\ 0.112\ 0.219$
> $0.332\ 0.233\ 0.621\ 0.179$
> $0.401\ 0.433\ 0.116\ 0.579$
> $0.279\ 0.511\ 0.62\ 0.222;$
> $M = 3\ 2\ 2\ 2;$
> $A = 9;$
> $enddata$
> end

The optimal solution of LPP (9.6) derived through LINGO-18 software is $x_{11} = 1$, $x_{12} = 1$, $x_{14} = 1$, $x_{21} = 1$, $x_{23} = 1$, $x_{32} = 1$, $x_{34} = 1$, $x_{42} = 1$, $x_{43} = 1$ with maximum Z = 4.119. That means vendor 1 should be supplying products 1, 2, & 4, vendor 2 should be supplying products 1 & 3, vendor 3 should be supplying products 2 & 4, and vendor 4 should be supplying products 2 & 3 respectively. ∎

9.5 Diet Optimization Problem

One more important category in which linear programming has been applied is "Diet Problem." Every human being has its nutrient requirements such as vitamins, proteins, fats, and carbohydrates. Also, each type of food possesses different quantities of these nutrients, which are known. The cost of each type of foodstuff is also known. Therefore, linear programming will help to decide the diet plan at minimum cost and fulfils the necessary daily requirements of nutrients. Few examples of the application of LP in the diet are as follows:

Example 126 *John is a health-conscious businessman. He wants to plan his diet in such a manner that it fulfils his basic daily requirements of nutrients (i.e., Fats, Proteins, Vitamins, and Carbohydrates) and at the same time minimizes his budget. He has four types of food items, among which he has to choose. The per-unit yields of these food items are given in Table (9.8).*

Table 9.8: Per-Unit Yields & Cost of Food Items.

Food Type	Yield per unit				Cost/unit ($)
	Proteins	Fats	Carbohydrates	Vitamins	
A	2	1	7	4	10
B	3	2	3	8	5
C	7	7	6	3	15
D	7	5	5	2	25

The minimum daily requirement of proteins, fats, carbohydrates, and vitamins are 500, 150, 600, & 700 respectively. Formulate the problem as an LP model.

Solution   Decision Variables:
x_1- Number of units of food type A
x_2- Number of units of food type B
x_3- Number of units of food type C
x_4- Number of units of food type D

Objective Function:

$$\min C = 10x_1 + 5x_2 + 15x_3 + 25x_4$$

Constraint 1: Protein requirement

$$5x_1 + 3x_2 + 7x_3 + 5x_4 \geq 900$$

Constraint 2: Fat requirement

$$x_1 + 2x_2 + 7x_3 + 5x_4 \geq 150$$

Constraint 3: Carbohydrates requirement

$$7x_1 + 3x_2 + 6x_3 + 5x_4 \geq 600$$

Constraint 4: Vitamin requirement

$$4x_1 + 8x_2 + 3x_3 + 2x_4 \geq 700$$

Non-Negativity Restriction:

$$x_1, x_2, x_3, x_4 \geq 0$$

Therefore, the LP model of the problem is as follows:

$$\begin{aligned}
\min \ C &= 10x_1 + 5x_2 + 15x_3 + 25x_4 \\
\text{subject to } & 5x_1 + 3x_2 + 7x_3 + 5x_4 \geq 900 \\
& x_1 + 2x_2 + 7x_3 + 5x_4 \geq 150 \\
& 7x_1 + 3x_2 + 6x_3 + 5x_4 \geq 600 \\
& 4x_1 + 8x_2 + 3x_3 + 2x_4 \geq 700 \\
\text{and } & x_1, x_2, x_3, x_4 \geq 0
\end{aligned} \qquad (9.7)$$

LINGO-18 codes for problem (9.7)

$MODEL:$
$SETS:$
$product/1..4/ : C, x;$
$const/1..4/ : mat;$
$material(const, product) : a;$
$endsets$
$[objective] \min = @sum(product(j) : C(j) * x(j));$
$@for(const(i) : [Material\_constraints]$
$@sum(product(j) : a(i,j) * x(j)) >= mat(i));$
$Data:$
$C = 10, 5, 15, 25;$
$mat = 900, 150, 600, 700;$
$a = 5\ 3\ 7\ 5 \quad 1\ 2\ 7\ 5 \quad 7\ 3\ 6\ 5 \quad 4\ 8\ 3\ 2;$
$enddata$
end

The optimal solution of LPP (9.7) derived through LINGO-18 software is $x_1 = 0,\ x_2 = 300,\ x_3 = 0,\ x_4 = 0$ with minimum cost = $1500. ∎

Example 127 *A person involved in livestock farming wants to purchase a feed mix containing at least minimum nutritional requirements. He can add one or more of the three types of grains that contain different quantities of nutritional elements. The required data is presented in Table (9.9).*

Table 9.9: Data of Nutritional Ingredients.

		Nutritional Ingredients				Cost/unit weight ($)
		A	B	C	D	
	Grain 1	2	2	4	5	40
Weight per unit	Grain 2	2	1	5	20	35
	Grain 3	7	0	0	2	95
	Requirements	1,350	250	800	240	

He wants to purchase this mix at minimum cost. Formulate the above problem as a LP model.

Solution   Decision Variables:
x_1 - Number of units of Grain 1
x_2 - Number of units of Grain 2
x_3 - Number of units of Grain 3

Objective Function:

$$\min\ C = 40x_1 + 35x_2 + 95x_3$$

Constraint 1: Requirement of nutritional ingredient A

$$2x_1 + 2x_2 + 7x_3 \geq 1,350$$

Constraint 2: Requirement of nutritional ingredient B

$$2x_1 + x_2 \geq 250$$

Constraint 3: Requirement of nutritional ingredient C

$$4x_1 + 5x_2 \geq 800$$

Constraint 4: Requirement of nutritional ingredient D

$$5x_1 + 20x_2 + 2x_3 \geq 240$$

Non-Negativity Restriction:

$$x_1, x_2, x_3 \geq 0$$

Therefore, the LP model of the problem is as follows:

$$\left.\begin{aligned}
\min\ C &= 40x_1 + 35x_2 + 95x_3 \\
\text{subject to}\ & 2x_1 + 2x_2 + 7x_3 \geq 1,350 \\
& 2x_1 + x_2 \geq 250 \\
& 4x_1 + 5x_2 \geq 800 \\
& 5x_1 + 20x_2 + 2x_3 \geq 240 \\
\text{and}\ & x_1, x_2, x_3 \geq 0
\end{aligned}\right\} \quad (9.8)$$

> **LINGO-18 codes for problem** (9.8)
> $MODEL:$
> $SETS:$
> $product/1..3/:C,x;$
> $const/1..4/:mat;$
> $material(const,product):a;$
> $endsets$
> $[objective]min = @sum(product(j):C(j)*x(j));$
> $@for(const(i):[Material\_constraints]$
> $@sum(product(j):a(i,j)*x(j)) >= mat(i));$
> $Data:$
> $C = 40,35,95;$
> $mat = 1350,250,800,240;$
> $a = 2\ 2\ 7\quad 2\ 1\ 0\quad 4\ 5\ 0\quad 5\ 20\ 2;$
> $enddata$
> end

The optimal solution of LPP (9.8) derived through LINGO-18 software is $x_1 = 75$, $x_2 = 100$, $x_3 = 142.86$ with minimum cost=$20071.43. ∎

9.6 Operations Management Optimization Problems

Optimization is used in operations management problems to aid in decision-making about product mix, production scheduling, staffing, inventory control, capacity planning, and other issues. In production scheduling, linear optimization plays an important role. Usually, a production scheduling problem has the objective of establishing an efficient, low-cost production schedule for one or more products over several periods under the constraints, including limitations on production capacity, labor capacity, storage space, and more. Some examples are as follows:

Example 128 *A company manufactures four types of products, say A, B, C, and D. Each product is processed on two machines X and Y. Product A takes 2 hours on machine X, and 3 hours on machine Y, product B takes 2 hours on X, 2 hours on Y, product C takes 3 hours on X, 1 hour on Y, while product D takes 4 hours on X and 5 hours on Y respectively. The total time available per month on each machine is 1,000 and 1,500 hours. The maximum market demand for product B is 500 units. Estimated profits on each product are $20, $30, $24, and $15, respectively. Formulate the LP model and derive the optimal solution.*

Solution   Decision Variables:
x_1- Number of units of product type A
x_2- Number of units of product type B

x_3- Number of units of product type C
x_4- Number of units of product type D

Objective Function:

$$\max\ Z = 20x_1 + 30x_2 + 24x_3 + 15x_4$$

Constraint 1: Processing time restriction on machine X

$$2x_1 + 2x_2 + 3x_3 + 4x_4 \leq 1,000$$

Constraint 2: Processing time restriction on machine Y

$$3x_1 + 2x_2 + x_3 + 5x_4 \leq 1,500$$

Constraint 3: Market restriction

$$x_2 \leq 500$$

Non-Negativity Restriction:

$$x_1, x_2, x_3, x_4 \geq 0$$

Therefore, the LP model of the problem is as follows:

$$\left. \begin{aligned} \max\ Z &= 20x_1 + 30x_2 + 24x_3 + 15x_4 \\ \text{subject to}\ & 2x_1 + 2x_2 + 3x_3 + 4x_4 \leq 1,000 \\ & 3x_1 + 2x_2 + x_3 + 5x_4 \leq 1,500 \\ & x_2 \leq 500 \\ \text{and}\ & x_1, x_2, x_3, x_4 \geq 0 \end{aligned} \right\} \quad (9.9)$$

LINGO-18 codes for problem (9.9)

```
MODEL:
SETS:
product/1..4/ : P,x;
const/1..3/ : mat;
material(const, product) : a;
endsets
[objective]max = @sum(product(j) : P(j)*x(j));
@for(const(i) : [Material_constraints]
@sum(product(j) : a(i,j)*x(j)) <= mat(i));
Data;
P = 20, 30, 24, 15;
mat = 1000, 1500, 500;
a = 2 2 3 4  3 2 1 5  0 1 0 0;
enddata
end
```

The optimal solution of LPP (9.9) derived through LINGO-18 software is $x_1 = 0$, $x_2 = 500$, $x_3 = 0$, $x_4 = 0$ with maximum profit=$15000. ∎

Example 129 *The management of XYZ hospital wants to determine the optimal number of nurses to be employed so that in each period, they have a sufficient number of nurses. The daily requirement of nurses in each period is presented in Table (9.10).*

Table 9.10: Daily Requirement of Nurses.

Period	Clock Time (24 hours day)	Minimum number of required nurses
1	6 AM-10 AM	2
2	10 AM-2 PM	7
3	2 PM-6 PM	15
4	6 PM-10 PM	8
5	10 PM-2 AM	20
6	2 AM-6 AM	6

The nurses have to report to the hospital at the beginning of the period and work for eight consecutive hours. Formulate the LP model to determine the minimum number of nurses to be employed.

Solution   Decision Variables:
x_1- Number of nurses in period 1
x_2- Number of nurses in period 2
x_3- Number of nurses in period 3
x_4- Number of nurses in period 4
x_5- Number of nurses in period 5
x_6- Number of nurses in period 6

Objective Function:

$$\min \ Z = x_1 + x_2 + x_3 + x_4 + x_5 + x_6$$

Constraint 1: Required number of nurses in period 2

$$x_1 + x_2 \geq 7$$

Constraint 2: Required number of nurses in period 3

$$x_2 + x_3 \geq 15$$

Constraint 3: Required number of nurses in period 4

$$x_3 + x_4 \geq 8$$

Constraint 4: Required number of nurses in period 5

$$x_4 + x_5 \geq 20$$

Constraint 5: Required number of nurses in period 6

$$x_5 + x_6 \geq 6$$

Constraint 6: Required number of nurses in period 1

$$x_6 + x_1 \geq 2$$

Non-Negativity Restriction:

$$x_1, x_2, x_3, x_4, x_5, x_6 \geq 0$$

Therefore, the LP model of the problem is as follows:

$$\left. \begin{aligned} \min \ Z &= x_1 + x_2 + x_3 + x_4 + x_5 + x_6 \\ \text{subject to} \ \ x_1 + x_2 &\geq 7 \\ x_2 + x_3 &\geq 15 \\ x_3 + x_4 &\geq 8 \\ x_4 + x_5 &\geq 20 \\ x_5 + x_6 &\geq 6 \\ x_6 + x_1 &\geq 2 \\ \text{and} \ \ x_1, x_2, x_3, x_4, x_5, x_6 &\geq 0 \end{aligned} \right\} \quad (9.10)$$

LINGO-18 codes for problem (9.10)

$MODEL:$
$SETS:$
$product/1..6/:x;$
$const/1..6/:mat;$
$material(const, product):a;$
$endsets$
$[objective]min = @sum(product(j):x(j));$
$@for(const(i):[Material\_constraints]$
$@sum(product(j):a(i,j)*x(j)) >= mat(i));$
$Data:$
$mat = 7,15,8,20,6,2;$
$a = 1\ 1\ 0\ 0\ 0\ 0\ \ 0\ 1\ 1\ 0\ 0\ 0\ \ 0\ 0\ 1\ 1\ 0\ 0\ \ 0\ 0\ 0\ 1\ 1\ 0\ \ 0\ 0\ 0\ 0\ 1\ 1\ \ 1\ 0\ 0\ 0\ 0\ 1;$
$enddata$
end

The optimal solution of LPP (9.10) derived through LINGO-18 software is $x_1 = 2$, $x_2 = 15$, $x_3 = 0$, $x_4 = 14$, $x_5 = 6$, $x_6 = 0$ with minimum nurses = 37. ∎

9.7 Transportation Optimization Problem

Suppose a Wuro-Dole (WD) Cement Company has six warehouses across a state each having a capacity of 80, 65, 58, 37, 73, and 52 tonnes and supplying to eight different cities each having a demand of 55, 37, 42, 52, 41, 22, 43, and 37 tones respectively. The supply of each warehouse cannot be exceeded during the month, and each city's demand must be satisfied. WD wants to determine how many tonnes of cement to ship from each warehouse to each city to minimize the total shipping cost. The unit cost of shipment from warehouse i to city j is given in the matrix below, and network flow is shown in Fig. (9.1).

$$\begin{bmatrix} Cities & 1 & 2 & 3 & 4 & 5 & 6 & 7 & 8 \\ Warehouse1 & 7 & 5 & 6 & 7 & 4 & 2 & 5 & 9 \\ Warehouse2 & 4 & 9 & 5 & 3 & 8 & 5 & 8 & 2 \\ Warehouse3 & 5 & 2 & 1 & 5 & 7 & 7 & 3 & 3 \\ Warehouse4 & 6 & 6 & 7 & 1 & 9 & 2 & 7 & 1 \\ Warehouse5 & 2 & 3 & 9 & 5 & 7 & 2 & 6 & 5 \\ Warehouse6 & 5 & 5 & 6 & 5 & 8 & 1 & 4 & 2 \end{bmatrix}$$

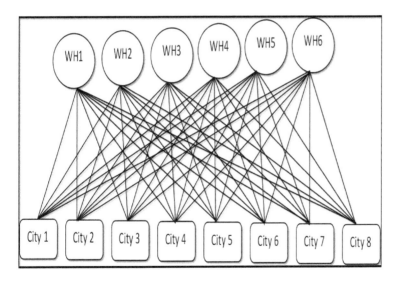

Figure 9.1: Cement Shipment Network Flow Diagram.

Since each warehouse can ship to any of the cities, there will be $6 \times 8 = 48$ possible routes as in Fig. (9.1). Let

x_{ij} = The quantity of cement (in tonnes) to be ship from warehouse i to demand city j.
C_{ij} = The unit cost of transhipment from warehouse i to demand city j in rupees.

S_i = The amount of cement (in tonnes) available at warehouse i.
D_j = The quantity of cement (in tonnes) demanded at city j.

Then, the mathematical formulation can be done as:

Objective Function:

$$\min\ C = \sum_{i=}^{6} C_{ij}x_{ij}$$

Constraint 1: Supply Constraints

$$\sum_{j=1}^{8} C_{ij}x_{ij} = S_i \ \forall\ i = 1, 2, \ldots, 6$$

Constraint 2: Demand Constraints

$$\sum_{i=1}^{6} C_{ij}x_{ij} = D_j \ \forall\ j = 1, 2, \ldots, 8$$

Non-Negativity Restriction:

$$x_{ij} \geq 0$$

Therefore, the LP model of the problem is as follows:

$$\left.\begin{array}{c} \min\ C = \sum_{i=}^{6} C_{ij}x_{ij} \\ \text{subject to}\ \sum_{j=1}^{8} C_{ij}x_{ij} = S_i \\ \sum_{i=1}^{6} C_{ij}x_{ij} = D_j \\ \text{and}\ x_{ij} \geq 0 \end{array}\right\} \quad (9.11)$$

> **LINGO-18 codes for problem** (9.11)
> MODEL:
> !Cement Transportation problem from six different Warehouses to eight different Cities;
> SETS:
> WAREHOUSES : CAPACITY;
> CITIES : DEMAND;
> LINKS(WAREHOUSES,CITIES) : COST,QUANTITY;
> ENDSETS
> !Here is the data;
> DATA:
> !set members;
> WAREHOUSES = WH1 WH2 WH3 WH4 WH5 WH6;
> CITIES = C1 C2 C3 C4 C5 C6 C7 C8;
> !attribute values;
> CAPACITY = 80 65 58 37 73 52;
> DEMAND = 55 37 42 52 41 22 43 37;
> COST = 7 5 6 7 4 2 5 9
> 4 9 5 3 8 5 8 2
> 5 2 1 5 7 7 3 3
> 6 6 7 1 9 2 7 1
> 2 3 9 5 7 2 6 5
> 5 5 6 5 8 1 4 2;
> ENDDATA
> !The objective function is to minimize the total cost of transportation;
> MIN = @SUM(LINKS(I,J) :
> COST(I,J) ∗ QUANTITY(I,J));
> !The demand constraints;
> @FOR(CITIES(J) :
> @SUM(WAREHOUSES(I) : QUANTITY(I,J)) = DEMAND(J));
> !The supply constraints;
> @FOR(WAREHOUSES(I) : @SUM(CITIES(J) : QUANTITY(I,J)) <= CAPACITY(I));
> END

The optimal solution of LPP (9.11) derived through LINGO-18 software is that the minimum cost = $780.

9.8 Assignment Optimization Problem

A manager has four jobs to be completed by four machines. One job must be assigned to each machine to complete. The time required for each machine for completing each job is given in Table (9.11). The manager wants to minimize the total setup time needed to complete the four jobs.

Table 9.11: Time Requirement Data.

	Time in Hours			
Machine	Job 1	Job 2	Job 3	Job 4
1	10	4	6	5
2	2	8	5	4
3	6	7	2	10
4	2	5	7	9

To determine which machine should be assigned to each job, we define decision variable for $i, j = 1, 2, 3, 4$

$x_{ij} = 1$ if machine i is assigned to job j
$x_{ij} = 0$ if machine i is not assigned to job j

Then, the LP model of the problem is as follows:

$$\begin{aligned}
\min \ z = &10x_{11} + 4x_{12} + 6x_{13} + 5x_{14} + 2x_{21} + 8x_{22} + 5x_{23} + 4x_{24} \\
& + 6x_{31} + 7x_{32} + 2x_{33} + 10x_{34} + 2x_{41} + 5x_{42} + 7x_{43} + 9x_{44} \\
\text{subject to } & x_{11} + x_{12} + x_{13} + x_{14} = 1 \quad \text{(Machine constraints)} \\
& x_{21} + x_{22} + x_{23} + x_{24} = 1 \\
& x_{31} + x_{32} + x_{33} + x_{34} = 1 \\
& x_{41} + x_{42} + x_{43} + x_{44} = 1 \\
& x_{11} + x_{21} + x_{31} + x_{41} = 1 \quad \text{(Job constraints)} \\
& x_{12} + x_{22} + x_{32} + x_{42} = 1 \\
& x_{13} + x_{23} + x_{33} + x_{43} = 1 \\
& x_{14} + x_{24} + x_{34} + x_{44} = 1 \\
& x_{ij} = 0/1
\end{aligned} \quad (9.12)$$

LINGO-18 codes for problem (9.12)

$MODEL:$
$SETS:$
$MACHINES/1..4/;$
$JOBS/1..4/;$
$LINKS(MACHINES, JOBS): COST, ASSIGN;$
$ENDSETS$
!Here is the data;
$DATA:$
$COST = 10, 4, 6, 5,$
$2, 8, 5, 4,$
$6, 7, 2, 10,$
$2, 5, 7, 9;$
$ENDDATA$
!The objective function is to minimize the total setup time needed to complete the four task;
$MIN = @SUM(LINKS: COST * ASSIGN);$
$@FOR(MACHINES(I):$
$@SUM(JOBS(J): ASSIGN(I,J)) < 1);$
$@FOR(JOBS(J): @SUM(MACHINES(I): ASSIGN(I,J)) > 1);$
END

The optimal solution of LPP (9.12) derived through LINGO-18 software is $x_{12} = x_{24} = x_{33} = x_{41} = 1$ with the minimum assignment cost = \$12.

References

[1] RE Bellman and LA Zadeh. Decision-making in a fuzzy environment. *Management Science*, 17(4):B–141, 1970.

[2] Asoke Kumar Bhunia, Laxminarayan Sahoo, and Ali Akbar Shaikh. Advanced optimization and operations research, 2019.

[3] John R Birge and Francois Louveaux. *Introduction to Stochastic Programming*. Springer Science & Business Media, 2011.

[4] Rainer E Burkard. Nonlinear programming: Theory and algorithms: Ms bazaraa and cm shetty wiley, chichester, 1979, 1980.

[5] Alberto Cambini and Laura Martein. *Generalized Convexity and Optimization: Theory and Applications*, volume 616. Springer Science & Business Media, 2008.

[6] Vira Chankong and Yacov Y Haimes. *Multiobjective Decision Making: Theory and Methodology*. Courier Dover Publications, 2008.

[7] A Charnes and WW Cooper. Multicopy traffic network models. *Theory of Traffic Flow*, 85:85–96, 1961.

[8] Abraham Charnes and William W Cooper. Chance-constrained programming. *Management Science*, 6(1):73–79, 1959.

[9] Abraham Charnes and William W Cooper. Deterministic equivalents for optimizing and satisficing under chance constraints. *Operations Research*, 11(1):18–39, 1963.

[10] Abraham Charnes, William W Cooper, and Robert O Ferguson. Optimal estimation of executive compensation by linear programming. *Management Science*, 1(2):138–151, 1955.

[11] Abraham Charnes, William W Cooper, and Bob Mellon. Blending aviation gasolines–a study in programming interdependent activities in an integrated oil company. *Econometrica: Journal of the Econometric Society*, pages 135–159, 1952.

[12] Michael Alan Howarth Dempster, Michael Alan Howarth Dempster, et al. *Stochastic Programming*. Academic Pr, 1980.

[13] Didier J Dubois. *Fuzzy Sets and Systems: Theory and Applications*, volume 144. Academic press, 1980.

[14] HA Eiselt, Carl-Louis Sandblom, et al. Nonlinear optimization. *International Series in Operations Research and Management Science*, 2019.

[15] Thomas Finney. Calculus and analytical geometry, 1989.

[16] Juergen Guddat, F Guerra Vasquez, Klaus Tammer, and Klaus Wendler. Multiobjective and stochastic optimization based on parametric optimization. *Mathematical Research*, 26, 1985.

[17] N Gupta and JK Sharma. Fuzzy multi-objective programming problem for revenue management in food industry. *Journal of Revenue and Pricing Management*, 2020.

[18] Petr Hájek. *Metamathematics of Fuzzy Logic*, volume 4. Springer Science & Business Media, 2013.

[19] C-L Hwang and Abu Syed Md Masud. *Multiple Objective Decision Making—Methods and Applications: A State-of-the-art Survey*, volume 164. Springer Science & Business Media, 2012.

[20] JP Ignizio. Linear programming in single and multiple objective systems, 1982.

[21] George Klir and Bo Yuan. *Fuzzy Sets and Fuzzy Logic*, volume 4. Prentice Hall New Jersey, 1995.

[22] Sang M Lee et al. *Goal Programming for Decision Analysis*. Auerbach Publishers Philadelphia, 1972.

[23] William S Meisel, JL Cochrane, and M Zeleny. Tradeoff decision in multiple criteria decision making. *Multiple Criteria Decision Making*, pages 461–476, 1973.

[24] Kaisa Miettinen. *Nonlinear Multiobjective Optimization*, volume 12. Springer Science & Business Media, 2012.

[25] Umar Muhammad Modibbo, Irfan Ali, and Aquil Ahmed. Multi-objective optimization modelling for analysing sustainable development goals of nigeria: Agenda 2030. *Environment, Development and Sustainability*, pages 1–35, 2020.

[26] Mohammad Asim Nomani, Irfan Ali, Armin Fügenschuh, and Aquil Ahmed. A fuzzy goal programming approach to analyse sustainable development goals of india. *Applied Economics Letters*, 24(7):443–447, 2017.

[27] András Prékopa. *Stochastic Programming*, volume 324. Springer Science & Business Media, 2013.

[28] Singiresu S Rao. *Engineering Optimization: Theory and Practice*. John Wiley & Sons, 2019.

[29] SS Rao. *Optimization: Theory and Applications*. Wiley Eastern Limited, 1979.

[30] SS RAO III. Nonlinear programming iii: Constrained optimization techniques. *Engineering Optimization: Theory and Pratice. 4^a. ed. Hoboken: John Wiley & Sons, Inc*, pages 422–428, 2009.

[31] Jati K Sen-Gupta. Stochastic programming: Methods and applications. 1972.

[32] SM Sinha. *Mathematical Programming: Theory and Methods*. Elsevier, 2005.

[33] Steven Vajda. *Probabilistic Programming*. Academic Press, 1972.

[34] Chankong Vira and Yacov Y Haimes. Multiobjective decision making: theory and methodology. *Noth-Holland Series in System Science and Engineering*, pages 62–109, 1983.

[35] Ronald R Yager and Lotfi A Zadeh. *An Introduction to Fuzzy Logic Applications in Intelligent Systems*, volume 165. Springer Science & Business Media, 2012.

[36] Lotfi A Zadeh. Fuzzy sets. *Information and Control*, 8(3):338–353, 1965.

[37] M Zeleny. Linear multiobjective programming springer-verlag. *Berlin, New York*, 1974.

[38] Milan Zeleny. *Multiple Criteria Decision Making Kyoto 1975*, volume 123. Springer Science & Business Media, 2012.

[39] HJ Zimmermann. Description and optimization of fuzzy systems. *International Journal of General Systems*, 2(1):209–215, 1976.

Index

A

Absolute Extreme Values 13
Absolute Value Function 6
Active 44, 107, 116, 132–134
Additivity 69
Advertisement 93, 206, 207, 209
Affine Sets 47
Algebra 22–24, 31
Algorithm 53, 124, 126, 129, 132, 133, 135, 136, 138, 172, 173, 183, 193, 194, 196–198
Allocation 40, 74, 100, 101, 189, 192, 193, 200, 206, 208, 209
Alpha-Cut 149, 150, 152, 155–157, 159, 160, 198
Alternate 83
Angle 3
Anti-Ideal Point 180, 183
Application 36, 46, 76, 78, 177, 200, 206, 217
Arbitrarily 10, 116, 126
Area 77, 78, 152–155, 200, 212
Arithmetic Operations 170, 175
Artificial 83, 88, 89, 124, 125
Assets 200, 212
Assignment 43, 76–78, 212, 214, 227, 229
Associative 24
Availability 69, 70, 73, 75, 95, 96, 109, 162, 189, 198, 206
Average Rate of Change 8, 12

B

Balanced 74, 75, 178
Basic solution 71, 123, 125
Basis 86–88, 90–92, 102, 111–113, 123, 125, 127
Beale 123, 126, 127, 129
Best Lower Bound 174, 175
Bit 36
Branch-and-Bound 78
Branching 43
Budget 76, 176, 206–210, 218
Business 200, 214

C

Calculus 2, 126
Capacity 40, 41, 43, 44, 92–94, 97–99, 114, 221, 225, 227
Capital budgeting 75, 76, 200
Carbohydrates 217, 218
Cardinality 145, 146
Center of Sums 152, 153
Certainty 69
Chance 162, 163, 165, 167, 169
Chief marketing 208
Circle 2, 5, 7
Classical 141, 142
Closure 116
Codes 37, 41, 92–94, 96–101, 113, 114, 119–122, 129–131, 139, 140, 157–160, 168–170, 174, 175, 190–196, 198, 199, 203, 205, 208, 211, 214, 217, 219, 221, 222, 224, 227, 229
Codomain 143
Coefficient 70, 83, 85, 88, 102, 103, 110, 156–158, 167, 171–173, 182, 186, 196
Cofactor 28

Column 22–24, 26–29, 31, 34, 85–87, 90, 91, 102, 111–113, 133, 213
Column Rank 26
Command 39, 44
Commutative 24
Complementary 106, 107, 124
Complex 22, 40, 68, 106
Components 31, 61, 69, 84, 85, 102, 104, 115, 116, 121, 127, 133
Composite 5
Compromise 199
Concave 16, 17, 19, 20, 49–57, 61, 119, 123, 126, 127, 131, 143
Concavity 17, 18, 50, 57
Cone 58
Conflicting 176, 177, 179, 182
Constants 6, 9, 10, 14, 28, 30, 42, 43, 58, 60, 69, 110, 162, 183
Constrained 46, 47, 55, 57, 59–61, 77, 132, 162, 165
Constraint 39–44, 46, 47, 52, 53, 55, 60, 61, 68–73, 75, 76, 78, 83, 84, 86, 88, 89, 92–107, 109, 110, 112, 114–122, 124, 126, 127, 131–138, 141, 156, 158, 162–166, 168–173, 177, 182–186, 188, 192–194, 196, 201–211, 213–215, 217–224, 226–228
Continuity 8, 9, 144
Continuous 8–10, 12–16, 59, 69, 78, 143, 145, 154
Contour 54, 57
Convergence 9, 138
Convex 16, 19, 32, 48–61, 68, 71, 72, 83, 123, 126, 131–133, 136, 138–140, 143, 181
Coordinate 2, 3, 8, 78–83
Cost 13, 39, 40–42, 68–70, 72, 74–77, 95, 96, 98, 99, 101–103, 109, 111, 119–121, 177, 189–192, 198, 199, 206, 209, 210, 214, 217–221, 225, 227, 229
Covariance 161, 164, 165
Crisp 141–143, 148, 149, 152, 154–160, 196–198
Criterion 127–129, 138, 177, 181
Critical 14, 16, 19, 61–64, 66, 67
Critical Point 14, 16, 61–64, 66, 67
Current Basic Feasible 87, 90, 91, 112, 113
Curve 5, 6, 8, 10, 11, 14, 18

Curvilinear 10
Customer Reach 208
Cutting Plane 78, 136
Cycling 88

D

Daily Requirement 217, 218, 223
Dantzig 123
Decision Maker 170, 177, 180, 182–184, 186
Decision Variable 69, 70, 72, 77, 78, 83, 102, 115, 172, 177, 185, 187, 201, 204, 207, 209, 212, 215, 218, 220, 221, 223, 228
Decisions 46, 69, 70, 72, 77, 78, 83, 102, 115, 170, 172, 176, 177, 180–187, 196, 200, 201, 204, 207, 209, 212, 215, 218, 220, 221, 223, 228
Decreasing Function 15, 17
Defuzzification 152, 156, 197, 198
Degree 69, 142, 145, 186
Demand 41, 68, 77, 95, 96, 98–101, 188, 198, 221, 225–227
Derivative 10–13, 15–17, 20, 57, 61–66, 116
Desired Goal 190, 191
Desktop 37
Determinant 22, 26–29, 66
Deterministic 161, 162, 164–170
Deviational 173, 175, 177, 178, 182–184
Diagonal 23, 25, 28, 29
Diet 74, 75, 217, 218
Differentiable 12, 14, 15, 17, 20, 56, 57, 59, 65, 117, 118
Dimensional 31–33, 61, 83, 133, 187
Direct 38, 129–131, 139, 140, 174, 175, 184
Direction 31, 58, 65, 116, 117, 131, 132, 134
Directional Derivative 65, 116
Discrete 145, 154
Distance 2, 31, 32, 177
Distributive 24
Division 171
Domain 4–6, 8–10, 13, 14, 43, 44, 54, 56–59
Dual 39, 42, 102–112
Dual Price 39, 42
Duality 102, 103, 106, 107, 109, 111

Index ■ 235

E

Economic 109, 110
Editing 36
Efficient 78, 132, 178, 180, 181, 187, 197, 221
Eigenvalue 34, 35
Eigenvector 34, 35
Element 4, 22, 23–25, 28–30, 32, 58, 71, 86, 87, 90, 91, 111, 113, 141–143, 145, 149, 154, 183, 196, 219
Empty 31, 32, 70, 71
Entering Variable 87, 91, 113
Equality 32, 33, 42, 52, 78, 86, 89, 105, 106, 112, 117, 177, 187
Equation 2, 3, 6–8, 20, 27, 30, 31, 35, 40, 67, 72, 83, 88, 102–104, 123, 124, 126–128, 167, 177, 184
Euclidean 32
Exclamation 41
Expected Return 200, 201
Extreme 13, 14, 16, 19, 32, 49, 59, 66, 71, 72, 78–83

F

Facebook 208–210
Feasibility 107, 111, 211
Feasible Direction 116, 117, 131, 132
Feasible Region 60, 68, 77–83, 108, 127, 136, 179, 180
Feasible Solution 70–72, 83, 84, 86–88, 90, 91, 107–109, 111–113, 115, 116, 124–127, 132, 178, 186
Finance 68, 200, 212, 213
Finite 10, 12, 32, 43, 70, 71, 83, 88, 107, 109
Firm 75, 130, 214
Forbenlus 34
Formulation 68, 72–74, 76, 83, 110, 131, 171, 177, 182, 184, 226
Frontier 180, 181
Function 3–20, 31, 32, 39–44, 46–57, 59–72, 79–85, 88, 92–94, 96–99, 102, 103, 107, 109, 110, 114, 115, 117, 118, 122, 126–128, 131, 132, 141–147, 149–152, 154, 156, 157, 161, 163, 164, 166–168, 170–173, 175, 177, 178, 181–188, 190, 191, 195–199, 201, 203, 204, 205, 207, 209, 212, 213, 215, 218, 220, 222, 223, 226, 227, 229

Funnel Analysis 209
Fuzziness 196
Fuzzy 46, 141–160, 186–188, 195–199
Fuzzy Decision 196
Fuzzy Number 143, 151, 152, 155–160, 196–198

G

Global 13, 49, 50, 53, 56, 57, 59–64, 93–101, 114, 127, 129–132, 139, 140, 157–160, 169, 190, 191
Goal Programming 177, 182, 183, 185, 186, 188, 190, 191, 193, 195
Goals 46, 173, 175–178, 182–188, 190, 191, 193–197
Golden Section 59
Graph 5–8, 12, 17, 18, 20, 50, 78–83, 178
Graphical 78–83, 91, 156
Gravity 152, 154

H

Health-Conscious 218
Height 143
Hessian 56, 65–67
Hospital 223
Hull 50, 58
Human Resource 68, 200, 212
Hyperplane 52, 56, 58, 59, 136

I

Ideal Point 179, 180, 183
Ideal Solution 178, 197
Identity Matrix 23
Inclination 3
Increasing 15, 17, 54, 56, 127, 212
Increasing Function 15, 17
Independent 9, 10, 26, 27, 84, 168
Independently 162, 165, 169
Inequalities 68, 104, 105, 163
Inequality 2, 7, 32, 47, 50, 52, 53, 57, 86, 89, 112, 163, 176, 178
Infeasible 52, 70, 83, 109, 111, 179, 211
Inflection 18, 60–62, 64, 66
Initial Basic Feasible 86, 90, 112
Input Data 42
Install 37
Instantaneous Rate 12
Integer 36, 43, 44, 76–78, 119–121, 133

Intersection 48–50, 68, 144, 147
Interval 4, 8, 12–15, 17, 56, 59, 61, 141, 142, 170–173, 175
Interval Numbers 170–173
Inverse 25, 29
Investment 68, 75, 76, 200–204
Iteration 83, 88, 111, 113, 126, 128, 132, 133, 135–138
Iterative 83

J

Job 76, 77, 100, 101, 227–229

K

Kelly 135, 136
Kernel 143
Key Column 86, 87, 90, 91, 112, 113
Keys 36, 44, 86, 87, 90, 91, 112, 113, 214
Kuhn-Tucker 118, 123, 126

L

Labor 43, 44, 221
Lagrange 117, 118
Lagrange Multiplier 117, 118
Language 36, 40
Law of Contradiction 148
Law of Excluded 148
Leaving 87, 90, 91, 103, 112, 113
Lemke 123
Level Set 143, 149, 150
Lexicographic Goal Programming 185, 191
Limit 10–13, 43, 44, 69, 136, 163, 171, 177, 187, 197, 198, 207, 215
Limitation 46, 69, 119, 121, 182, 215, 221
Line 2, 3, 5, 8, 10, 11, 18, 20, 21, 32, 43, 44, 47, 50, 54, 78, 178
Linear 27, 30, 32, 36, 46, 47, 49, 50, 52–55, 58, 60, 68, 70, 72, 83, 115, 119–123, 125, 126, 132, 135, 136, 156, 161, 164, 171, 173, 177, 183, 196, 197, 199, 206, 209, 212, 217, 221
Linear Dependence 27
Linear Membership 196, 197, 199
Linear Programming 68, 70, 115, 125, 183, 206, 209, 212, 217
Linearization 20, 21
Linearize 136, 137

LINGO 36–43, 91–101, 113, 114, 119–122, 129–131, 138–140, 156–160, 167–170, 173–175, 184, 187–199, 203, 205, 206, 208, 211, 214, 217, 219, 221–224, 227, 229
LINGO-18 36, 91–101, 113, 114, 119–122, 129–131, 138–140, 156–160, 167–170, 173–175, 187–199, 203, 205, 206, 208, 211, 214, 217, 219, 221–224, 227, 229
Linguistic 142
Lipschitz Constant 9, 10
Livestock 219
Local 13, 16–18, 20, 21, 49, 50, 53, 56, 57, 59–62, 66, 127, 132, 168, 170
Local Extreme Values 13, 16
Local Minimum 13, 16, 49, 59, 61, 132
Logical Operators 42
Looping 40, 42, 43
Loss of Generality 84, 116, 135
Lower Triangular 29

M

Machine 76, 77, 100, 101, 109–111, 197, 198, 221, 222, 227–229
Machining 92
Magnitude 31, 184
Management 68, 212, 221, 223
Managers 109, 188, 193, 197, 200, 206, 214, 227, 228
Manufacturing 40, 97, 110, 197, 212
Mapping 4, 31
Marginal 13, 109
Material 40, 41, 68, 92–98, 114, 130, 193, 194, 203, 205, 208, 211, 212, 219, 221, 222, 224
Mathematical 47, 52, 68, 70, 76, 115, 122, 131, 141, 171, 184, 226
Matrices 25, 26, 30, 126
Matrix 22–30, 34, 43, 56, 65–67, 70, 84, 85, 102, 123, 125, 126, 134, 161, 164, 195, 212, 213, 225
Maxima 13, 17–21, 53, 59, 61–63, 66, 70
Maximum 13, 14, 16, 19, 20, 26, 34, 43, 50, 52, 54, 55, 59, 63, 64, 66, 67, 80, 81, 93–95, 97, 98, 100, 107, 109–112, 120, 122, 127, 152, 157–160, 163, 190–192, 197, 203, 206, 208, 212, 214, 215, 217, 221, 223

Mean 14, 15, 152, 161, 163, 164, 166–168
Mean Value 14, 15
Media 200, 206, 208, 209
Membership 141–147, 149–152, 154, 187, 188, 195–197, 199
Method 59, 60, 78–84, 86, 88, 89, 91, 111, 112, 123–127, 129, 132, 135, 136, 152–160, 181, 184, 191, 193, 197, 198, 211
Metric 31, 32
Minima 13, 18–21, 53, 56, 60–63, 66, 70
Minimum Ratio 87, 90, 112, 125
Minor 28, 66
Multi-Objective 176, 178, 181, 182, 184, 188, 195, 196, 198
Multiple 28, 46, 83, 176, 183, 196
Multiplication 23, 24, 171, 185
Mutual Funds 200–204

N

Nadir Point 180
n-Dimensional 31, 61, 83, 187
Negative Definite 54, 66, 67
New Element 86, 87, 90, 91, 113
New Window 39
Newton-Raphson 59
No Solution 85, 109, 111, 123, 127, 178
Non-Basic 85, 102, 103, 125–127
Non-Degenerate 71, 84, 88, 126
Non-Dominated 178, 181
Non-Linear 36, 46, 47, 115, 119–121, 123, 135, 136, 177, 183, 196
Non-Linear Programming 115, 183
Non-Negativity 71–73, 75, 123, 202, 204, 207, 210, 218, 220, 222, 224, 226
Non-Singular 25, 27, 29, 30, 102
Norm 32–34, 52
Normal 58, 143, 146, 161, 163, 164, 166, 168
Normal Distribution 161, 163
Normalized 134
Normally 161, 162, 164, 167, 169, 182
Normed Linear Spaces 32
Null Matrix 23
Numbers 4, 22, 23, 26, 27, 42, 61, 69, 72–75, 83, 84, 88, 89, 102, 104, 106, 110, 132, 136–138, 142, 143, 151, 152, 154–160, 170–173, 177, 193, 194, 196–198, 203, 206, 207, 214, 218, 220–224
Nurses 223, 224
Nutrient 74, 75, 217, 218
Nutritional Requirements 219

O

One Dimensional 133
Operations 44, 46, 48, 111, 143, 170, 175, 212–214, 221
Operations Management 221
Operators 41, 42
Optimal 40, 60, 61, 70, 72, 78, 83, 87, 88, 90, 91, 93–101, 103, 106, 107, 109–114, 116–122, 129, 133, 136, 157–160, 168–170, 173, 175, 176, 178–180, 183, 185, 190–193, 200, 203, 206, 208, 209, 211, 214, 217, 219, 221, 223, 224, 227, 229
Optimal Solution 61, 70, 72, 78, 83, 87, 88, 91, 99–101, 103, 107, 110, 111, 113, 114, 116–122, 129, 136, 157–160, 168–170, 173, 175, 178–180, 185, 190–193, 203, 206, 208, 211, 214, 217, 219, 221, 223, 224, 227, 229
Optimality 59, 61, 72, 85, 88, 90, 103, 107, 111, 128, 129, 183
Optimization 9, 22, 36, 45–47, 52, 53, 55, 57, 59–61, 68, 91, 115, 119–121, 129, 141, 156, 161, 168–171, 173, 175–177, 179–183, 187, 188, 194–197, 199, 200, 206, 212, 214, 217, 221, 225, 227
Optimum 40, 47, 59, 73, 74, 78–83, 85, 87, 91, 109, 111, 113, 126, 129–131, 135, 139, 140, 178, 181, 186, 189, 194, 198, 212
Ordinary 89
Orthogonal 26

P

Package 36, 68
Parabola 7, 8
Parabolic 7
Parameters 55, 69, 84, 104, 124, 141, 156, 159, 161, 165, 170–172, 188, 189, 197, 198
Pareto 178–181, 183, 186, 193
Pareto Front 181

Partial Derivatives 62–66
Penalty 89, 193
Phase-I 123, 125
Plane 2, 78, 136
Plant Capacities 94, 97
Plotted 79–83
Point-Slope 20
Polyhedron 83
Polynomial 9, 34
Portfolio 200, 203
Positive Definite 54, 56, 60, 66
Preemptive 185, 191
Preference Weights 214, 215
Primal 102–111
Print 42, 44, 208, 209
Print Media 209
Priority 43, 185
Probability 163
Product 23, 26, 28, 30–32, 43, 44, 69, 72–74, 86, 87, 90, 92–99, 102, 109, 110, 113, 114, 177, 197, 198, 203, 205, 206, 208, 211, 214, 215, 217, 219, 221, 222, 224
Product Mix 72–74, 92, 221
Production 13, 43, 44, 69, 73, 95, 96, 98, 99, 109–111, 130, 193, 206, 209, 212, 221
Production Cost 95, 96, 98
Profit 43, 44, 68–70, 72, 83, 92–95, 97, 98, 100, 103, 109–111, 177, 189, 193, 212, 221, 223
Programming 52, 68, 70, 76–78, 115, 122, 125, 129, 131, 133, 136, 138–140, 161, 162, 170, 177, 182, 183, 185, 186, 188, 190, 191, 193, 195, 197, 198, 206, 209, 212, 217
Progression 40
Prominent 203
Pseudo 57, 119

Q

Quadratic 8, 36, 52, 59, 60, 122, 123, 129
Quantity 31, 152, 225–227
Quasi Concave 53–55, 57, 119, 143

R

Radius 2, 7
Random 141, 161–169

Random Variables 161–169
Range 4, 5, 43, 44, 186, 206
Rank 23, 26, 27, 30, 102
Ranking 152, 185, 191
Raw Materials 40, 41, 130, 193
Real 4, 22, 23, 43, 46, 59, 61, 83, 84, 141, 143, 176, 183, 197
Real Value 4, 43
Rectangular 22
Reduced Cost 39, 42
Region 60, 68, 77–83, 108, 127, 136, 179, 180
Regret Value 213
Relative 18–20, 61, 102, 103, 145, 161, 167, 212
Relative Maxima 18, 19, 61
Relative Minimum 19
Reliability 119–122
Resource 68–70, 72–74, 83, 109, 114, 176, 182, 186, 200, 212
Restriction 43, 47, 72, 73, 75, 110, 117, 121, 123, 173, 202, 204, 206, 207, 210, 213, 216, 218, 220, 222, 224, 226
Restricts 4, 43
Revenue 189–192, 198
Risk 200, 201, 203, 204, 214
Rolle's Theorem 14
Rosen's Gradient Projection 132
Row 22, 23, 26–29, 31, 34, 86, 87, 90, 91, 111, 113, 213
Row Rank 26

S

Saddle Point 59, 61, 62, 65
Sales 93, 206
Scalar 23, 27, 28, 48, 145, 171, 181
Scare Resources 212
Scheduling 68, 221
Screen 39
Secant 8, 9, 12
Semi-Colon 42
Set 4, 30–32, 34, 36, 38, 40–43, 46–60, 62, 68, 70–72, 83–86, 92–94, 96–101, 107, 108, 111, 114, 116–118, 123, 131–133, 136, 138, 141–150, 152–156, 176, 181–184, 186, 187, 191, 196, 197, 203, 205, 208, 211, 214, 217, 219, 221, 222, 224, 227, 229
Setname 40

Index ■ 239

Shadow Price 42, 109, 111
Shifting 6
Shipping 225
Shortcut 44
Simplex 78, 83–91, 111, 112, 123–126
Singular 25–27, 29, 30, 102
Skew Symmetric 23, 26
Slack 39, 42, 83, 86, 89, 112, 126, 127
Slackness 106, 107, 124
Slope 2, 3, 8, 10–13, 20, 59
Software 1, 36, 37, 40, 68, 92–101, 114, 129, 130, 138–140, 156, 168–170, 184, 187, 189, 192–194, 197, 199, 203, 206, 208, 211, 214, 217, 219, 221, 223, 224, 227, 229
Solution 27, 30, 31, 34, 37–39, 42, 44, 55, 61, 68, 70–72, 78–103, 107–133, 136–140, 157–160, 168–170, 173, 175–181, 183, 185–187, 189–201, 203, 204, 206–209, 211, 212, 214, 215, 217–221, 223, 224, 227, 229
Solvers 36–39, 42
Spectrum 34
Square Matrices 25, 30
Standard 52, 70, 72, 78, 83, 84, 86, 89, 102–104, 112, 115, 122, 124, 129, 146, 147, 150, 161, 164, 167, 168, 182, 184
Standard Deviation 167, 168
Stationary 15, 59, 118
Stochastic 36, 46, 141, 161, 162, 168, 169
Strategy 177, 209
Strictly 48, 50, 51, 53–57, 61, 178
Strong Duality 111
Sub-Matrix 28, 85
Subnormal 143, 146
Sufficiency 119
Supplier 214
Surplus 39, 42, 83, 86, 126
Survey 120, 121, 206, 208
Symbols 123, 166, 167, 186, 187
Symmetric 5, 23, 26, 43, 123
Syntax Error 37, 38

T

Tableau 85–88, 90–92, 111–113, 125, 126
Tangent 3, 10, 11, 14, 18, 20, 21, 56, 57
Time 14, 23, 27, 68, 70, 76–78, 109–111, 121, 176, 177, 189, 197, 198, 206, 207, 209, 214, 218, 221–223, 227–229

Time Required 109, 121, 227
Tolerance 136, 187, 188, 197, 198
Toolbar 37, 39, 42
Total Return 76, 201, 203, 204, 206
Trace 25
Transhipment 225
Transportation 40, 41, 68, 77, 78, 225, 227
Transpose 24, 29–31
Trapezoidal 151–154
Triangular 28, 29, 151, 152, 198
Two-Phase 123, 125

U

Unaltered 134
Unbounded 70, 83, 85, 109, 127, 137
Uncertain 141, 157–159, 167
Unconstrained 46, 47, 60, 61, 133
Undefined 14, 18, 19
Underscore Character 42
Union 144
Univariate 59, 61
Universal 144–146, 148–150
Unrestricted 59, 103–106
Unwanted 173, 182, 183, 185
Upper Bound 10, 71, 143, 155, 156, 170, 172–174
Usable 131, 132
Utility 184

V

Vaguely 141
Vanishes 127
Variables 3, 10, 11, 30, 39, 42, 43, 59, 61, 66, 68–72, 77, 78, 83, 85–91, 102–107, 109–113, 115, 118, 124–128, 142, 152, 161–169, 172, 173, 175, 177, 178, 182–185, 187, 201, 204, 207, 209, 212, 215, 218, 220, 221, 223, 228
Variance 119–121, 161, 163–166
Variance-Covariance 161, 164
Variation 163, 168
Vector 31, 33, 34, 47, 58, 61, 65, 70, 71, 84, 85, 88, 102–104, 106, 111, 115, 118, 126, 135, 162, 177, 178, 182, 185–188
Vendor 214, 215, 217
Version 36, 37, 87, 91, 103, 156
Vertical 2, 3, 5, 7, 8, 11, 12
Vitamins 74, 217, 218

W

Warehouses 40, 41, 78, 98, 99, 225–227
Weak Duality 107, 109
Weak Pareto 178, 186, 193
Weighted-Sum Technique 181

Windows 36–39, 44
Wolf 123
Worst Upper Bound 174

Z

Zero-One 76–78, 183